Isoenzymes

Isoenzymes

D.W. MOSS

Professor of Clinical Enzymology
Royal Postgraduate Medical School
Hammersmith Hospital
London

London   New York

Chapman and Hall

First published in 1982 by
Chapman and Hall Ltd
11 New Fetter Lane, London EC4P 4EE
Published in the USA by Chapman and Hall
in association with Methuen, Inc.
733 Third Avenue, New York NY 10017

© 1982, Chapman and Hall Ltd

Printed in Great Britain by
J.W. Arrowsmith Ltd, Bristol

ISBN 0 412 22200 0

British Library Cataloguing in Publication Data

Moss, D.W.

Isoenzymes.
1. Isoenzymes
574.19′25   QP601
ISBN 0–412–22200–0

Contents

Preface

The increased interest in multiple forms of enzymes that began with the application of new methods of fractionation to preparations of enzymes and other proteins some 25 years ago led quickly to an appreciation that the existence of enzymes in multiple forms, or isoenzymes, is a general phenomenon. The results of pioneering studies and those which followed in the early years of isoenzyme research consisted, not surprisingly, mainly of descriptions of the existence and characteristics of heterogeneity in various enzyme systems. Summaries of these results were provided in books such as J.H. Wilkinson's *Isoenzymes*, the first edition of which appeared in 1965. Some clearer ideas of the nature of the phenomena had become apparent by the time that the second edition of *Isoenzymes* was called for in 1970, and a limited use of the word isoenzymes itself, to describe only certain of the various categories of enzyme multiplicity then recognized, was already being proposed. Nevertheless, a largely enzyme-by-enzyme organization of the contents of the book was still appropriate.

Considerable advances, both experimental and conceptual, were made in isoenzyme research in the 1970s, and in 1977 Professor Wilkinson suggested to the present author that these should be taken into account in a joint revision of *Isoenzymes*. Professor Wilkinson's untimely death put an end to this project and the present book is therefore the work of a single author. Those who were familiar with Henry Wilkinson's work in clinical enzymology will appreciate the extent of the loss thus sustained.

It has seemed appropriate in writing this book to discard a solely phenomenological approach and to try instead to bring out those generalizations concerning the occurrence, nature, properties and, where possible, functions of multiple forms of enzymes which seem to be justified by the results of research. These are illustrated by examples drawn almost entirely from animal, and especially human, enzyme systems on the basis of the author's greater familiarity with them. Readers whose interests lie mainly in the biochemistry of plants and micro-organisms may be disappointed by the limited attention that these categories of living matter have received; nevertheless, I hope that such readers will find some general principles of interest to them. The title *Isoenzymes* has been retained without qualification for this reason. The term 'isoenzymes' is also still widely used in an operational sense to describe any multiple forms of an enzyme, whatever their origins. This has provided a further reason for retaining the simple title, although descriptions of multiple forms of enzymes which do not fall within the current formal definition of isoenzymes are included in the book.

I thank those authors, editors and publishers indicated in the text who have given permission for the use of illustrations. I am greatly indebted to my collaborators for their part in my own experimental work on multiple forms of enzymes, and all authors will appreciate my debt to Mrs Brenda Salvage who prepared the typescript.

Donald Moss
July 1981

1 Multiple Forms of Enzymes and the Emergence of the Isoenzyme Concept

The virtually limitless spectrum of chemical reactions catalysed by enzymes – far wider than the range of reactions influenced by inorganic or synthetic catalysts – was recognized early in the history of enzymology to be due to the existence of an almost equally wide range of enzymes, each with a characteristic specificity. In the third edition of his textbook of chemistry, published in 1837, J.J. Berzelius considered two alternatives: that a few enzymes with wide specificity might be responsible for this great range of catalytic ability, or that many specific enzymes might exist. He regarded the latter possibility as more likely (Dixon, 1971). Although the association of a uniquely-specific enzyme with each catalysed reaction could not be made, in view of the discovery of enzymes specific for particular chemical groups or reactions, classification of enzymes in functional terms became, and has remained, the most useful and practicable system. For equally valid reasons, the main effort in the systematic study of enzymes became concentrated on factors which influence the rate of the catalysed reaction, since this approach offered the best prospect of understanding the nature of the catalytic process and its functional significance. The possession of a particular type of catalytic ability thus became the primary consideration in the selection of enzymes for the study of aspects of catalysis, as it did in enzyme classification.

The early history of enzymology provides many examples of differences of properties between functionally-similar enzymes from different sources, and as early as 1895 Emil Fischer had noted the need to specify the origin of an enzyme when describing its properties. As well as differences between analogous enzymes from such dissimilar but frequently-used sources as yeast and mammalian tissues (e.g. yeast and liver alcohol dehydrogenases) differences between enzymes with similar catalytic actions from various tissues of a single species were also recognized before 1950, as for example in the case of non-specific acid phosphatases from human prostatic and other tissues. However, successful attempts to demonstrate differences between enzymes of wide distribution in human tissues were few before this date, so that the weight of opinion was against the existence of organ-specific enzyme variants.

Even when several studies of multiple forms of an enzyme had accumulated, authors continued to find it necessary to go to considerable lengths to anticipate objections that their observations were the result of artefacts of the experimental techniques employed, or of ill-defined phenomena such as aggregation or association of a single enzyme with other components. For example, in discussing these possibilities in relation to their own, and earlier, results on the heterogeneity of horse-radish peroxidase, including seasonal variations in the relative amounts of different components and differences in their reactivity towards various substrates, Jermyn and Thomas (1954) note that 'the existence of multiple components in naturally occurring enzymes is far from being generally accepted'. The viewpoint of classical enzymology towards analogous enzymes from different sources was expressed in the first edition of the authoritative monograph by Dixon and Webb (1958) in the words:

> 'It is a remarkable fact that in general the catalytic properties, specificity, activity, affinities, etc., of a given enzyme vary little with the source. Although there may be slight physical differences in a given enzyme when it is produced by different cells they are usually unimportant, and the enzyme remains essentially the same enzyme'.

Concentration on functional rather than structural aspects of enzymes was reinforced at first by uncertainties about the chemical nature of enzymes and later, when the protein nature of enzymes was accepted, by the absence of methods for the isolation and analysis of proteins. Some early experimental studies on protein structure, such as those made possible by the ultracentrifuge, seemed to encourage speculation that proteins would prove to have repeating structural elements in common, and that the possibilities for structural variations between molecules would consequently be limited (Fruton, 1979). However, elucidation of the amino acid sequences of proteins in increasing numbers from the mid-1950s onwards demonstrated their individuality and disposed of theories of protein structure which predicted the repetition of common structural elements at the primary level. Studies of the characteristics of enzymic catalysis had by this time established the concept of the active centre, a relatively small region of the molecule at which attachment of the substrate takes place. Therefore, the possession of identical active centres could be expected to endow analogous enzyme molecules from different sources with their common catalytic properties, while allowing scope for variations in other properties through structural differences in catalytically-inactive regions of their molecules.

In some respects the recognition that the identity of each polypeptide chain is determined by its specific amino acid sequence, and therefore that not only this primary structure but also the three-dimensional secondary and tertiary structures which follow from it are characteristic of a particular protein, may appear to impose new restrictions on the possibility of structural variation between functionally-similar proteins. Furthermore, investigations of the relationship between structure and function (e.g., in the case of haemoglobin) have drawn attention to the functional importance of structural features distant from the primary substrate- or ligand-binding site, seeming further to reduce the extent to which protein structures can differ while retaining an overall similarity of function. However, this latter consideration itself adds a new dimension of interest to the search for variant forms of enzymes and other biologically-active proteins, since it increases the likelihood that the structural differences between them will be associated with functional differences, the nature and significance of which would not be apparent from the study of a single molecular species.

The emergence of a generalized concept of the existence of enzymes in multiple forms was dependent on the development of means for the separation and characterization of closely similar protein molecules, through which the prevalence of such multiple forms came to be recognized.

Although analytical techniques such as moving-boundary electrophoresis had brought to light the heterogeneity of certain purified enzymes, e.g. of crystalline lactate dehydrogenase from beef heart (Neilands, 1952), the improvement of separative methods based on differences in net molecular charge was responsible for the great increase in interest in enzyme heterogeneity from the middle 1950s onwards. Chromatography on substituted-cellulose ion-exchange materials was used in some early studies of the multiple forms of lactate dehydrogenase – an enzyme with a central position in the development of the isoenzyme concept – and this is still an important preparative technique in isoenzyme studies. However, the demonstration of the widespread occurrence of enzymes in multiple forms is due mainly to the application of techniques of zone electrophoresis, especially with starch gel as the supporting medium, and with the adaptation of histochemical methods to visualize the separated enzyme zones *in situ* (Hunter and Markert, 1957). The 'zymogram' technique, as it has been called, has remained the most useful single experimental method in studying the multiple forms of enzymes, especially in the detection of enzyme heterogeneity in tissue-extracts or blood serum for clinical purposes or when screening for enzyme polymorphisms in human or animal populations. Its importance can be gauged by the fact that electrophoretic mobility has become the most widely accepted property by which the multiple forms of an individual enzyme are designated, with components being assigned serial numbers in order of decreasing anodal mobility.

Besides the technique of zone electrophoresis, however, a wide range of methods for the separation and characterization of enzymes is regularly brought into use, comprising various forms of chromatography, electrophoresis and electro-focusing, studies of kinetic and immunological properties, selective inactivation by various agents, and structural analyses of differing degrees of completeness, with the ultimate aim of defining the differences between multiple enzyme forms in molecular terms (Moss, 1979).

The first generally accepted descriptive term for the existence of different molecular forms of proteins with the same enzymatic specificity was introduced by Markert and Møller (1959), who coined the word *isozymes* to describe this phenomenon. As is often the case with new coinages, the derivation of the word aroused some controversy, some authors preferring the spelling *isoenzymes*, and as a result both forms have survived and are used interchangeably.

When first introduced, the term isozymes (or isoenzymes) was not restricted to multiple forms of an enzyme existing within a particular biological context, e.g. a single species, but alternative or more restricted applications have also been suggested. Augustinsson (1961) proposed that multiple enzyme forms should be regarded as isoenzymes only when the differences between them involved little or no variation in the combination of enzyme and substrate. While this emphasizes the concept of an invariate active centre with the possibility of some variation in structure in other molecular regions, multiple forms of an enzyme which exhibit significant differences in their catalytic properties are now regarded as being particularly interesting. Isoenzymes have also been regarded as multiple forms of enzymes having a common tissue of origin, with the term 'heteroenzymes' suggested for the more general case of catalytically-similar enzymes found in different organs or species (Wieland and Pfleiderer, 1962). However, these attempts at more restricted definitions have not received wide acceptance.

The earliest uses of the term isoenzymes were also without implications as to the reasons for the existence of the multiple enzyme forms so described, although the problems posed for genetics by the multiplicity of proteins with a common activity were soon recognized (Markert and Møller, 1959). However, as the nature of some multiple forms of enzymes became clearer through genetic and structural studies, it became possible to define isoenzymes in terms of their genetic origins. According to the current recommendations of the Commission on Biological Nomenclature of IUPAC-IUB (1977), isoenzymes are defined as multiple molecular forms of an enzyme occurring within a single species, as a result of the presence of more than one structural gene. The multiple genes may be due to the presence of multiple gene loci or of multiple alleles. (The term 'allelozymes' is also used to denote isoenzymes deriving from allelic genes). Also included in this definition of isoenzymes are those multiple forms of enzymes which arise by the association of protein subunits that are themselves products of distinct structural genes.

Variant forms of enzymes which originate by post-genetic modifications of a single polypeptide chain, as in the conversion of inactive precursors of proteolytic enzymes to their active forms, are not regarded as isoenzymes, nor are the covalently-modified (e.g. phosphorylated or dephosphorylated) or conformationally-different forms in which certain enzymes may exist, and through which regulation of their activities is effected. Changes such as these and the enzyme forms with a more or less transient existence to which they give rise are not considered in this book. However, other stable multiple forms of enzymes which do not appear to be of genetic origin will be described, although in many cases their nature and significance are imperfectly understood.

An analogy can be drawn between the periodic table of the elements, drawn up originally on the basis of similarity of properties of elements in the same group, and the present systematic list of enzymes first proposed by the Enzyme Commission of the International Union of Biochemistry, which classifies enzymes according to the nature of the reaction which they catalyse*. In the way that isotopes of an element with different nuclear structures but common properties share the same position in the periodic table, isoenzymes catalysing the same reaction are subsumed

---

* Enzymes are referred to by their trivial names in the text of this book. Their corresponding Enzyme Commission numbers are given in the index.

under the same identifying number in the Enzyme Commission's list. However, the analogy is not an exact one. Isotopes of a given element all possess identical arrangements of their outer electron shells and consequently are identical in their chemical properties. Members of a particular set of isoenzymes are generally not completely identical in their catalytic properties, except in the nature of the reaction which they catalyse, and the extent of such functional differences gives rise to disagreement in some cases as to where a distinction should be drawn between sets of isoenzymes on the one hand, and groups of distinct but similar enzymes on the other.

The formal definition of isoenzymes now current, with the distinct genetic origins of multiple forms of enzymes as its basis, avoids the problem of specifying the degree of functional similarity which is to be expected in deciding whether the multiple forms in question should be classed as isoenzymes or not. Such difficulties are generally resolved by usage. Thus, although various proteolytic enzymes, such as trypsin and chymotrypsin, are functionally similar and are clearly of distinct genetic origins, they are not regarded as isoenzymes, or even as multiple enzyme forms, but as distinct enzymes, because of the otherwise marked differences between them. Similarly, non-specific acid and alkaline phosphatases display considerable similarities with regard to substrate specificity, but these two classes of enzymes are also not considered to be isoenzymic.

In some instances the conventions by which similar enzymes are regarded as distinct and are assigned individual numbers in the list of enzymes derive from the dates and circumstances of their discovery. For forty years a distinction has been drawn between enzymes capable of hydrolysing esters of choline, the acetylcholinesterase ('true' cholinesterase) characteristic of nervous tissue and the cholinesterase ('pseudo' cholinesterase) of serum, on the basis of their different but overlapping substrate specificities, although in other catalytic properties these enzymes are closely similar. Their independent genetic origins were demonstrated by the discovery of inherited variants of serum cholinesterase soon after the introduction of suxamethonium into anaesthetic practice in 1949. The relationships between these catalytically-similar but genetically-distinct forms therefore fall within the scope of the current definition of isoenzymes and it seems likely that, had their discovery and characterization taken place in more recent years, they might have been regarded as such, rather than as separate enzymes with consecutive numbers in the Enzyme Commission's list, as at present.

However, opinion is by no means unanimous on the isoenzymic status of more recently recognized enzymes with similar but not identical catalytic properties. Hexokinases which convert glucose to glucose-6-phosphate are widely distributed in mammalian tissues, and multiple forms of these enzymes are generally considered to be isoenzymes. A kinase present in the liver of some species is distinguished from this group by its more restricted substrate specificity and its higher Michaelis constant for glucose. This enzyme, referred to as glucokinase, has been given a separate identifying number, although many workers consider it to be a member of the hexokinase isoenzyme system (Purich *et al.*, 1973). Particularly difficult problems of classification arise with enzymes such as the non-specific esterases, which exist in numerous multiple forms in many species. Individual forms can be distinguished on the basis of their relative specificites for various synthetic substrates, but groups of such esterases are in some cases encoded by structural genes which are closely linked on a single chromosome, suggesting the common evolutionary origin thought to be characteristic of isoenzymes.

Just as in older studies the absence of a general awareness of the existence of variants of a single enzyme caused each discovery of heterogeneity of a particular catalytic property (for example, in a tissue extract) to be seen as evidence for the existence of distinct and unrelated enzymes, current acceptance of the isoenzyme concept may predispose enzymologists to group together under this description catalytic activities, which on closer examination, are indeed found to be manifestations of the presence of distinct enzymes. An example of this tendency is provided by a minor component of tryosine aminotransferase activity found in the cytoplasm of rat liver. At first regarded as an isoenzyme of the main tyrosine aminotransferase of this tissue, subsequent investigation showed the minor activity to be due to aspartate aminotransferase, an enzyme with quite distinct properties (Spencer and Gelehrter, 1974).

In spite of problems of definition, the concepts embodied in terms such as 'isoenzymes', or even the less restrictive 'multiple molecular forms of enzymes', are valuable in directing attention to features of enzyme evolution, structure and function from which significant generalizations can be inferred. Some of the generalizations which have already emerged as a result of the stimulus given to enzyme research by the isoenzyme concept are outlined in the following chapters.

2 Origins and Structures of Multiple Forms of Enzymes

The definition of isoenzymes as the products of distinct structural genes implies that those multiple enzyme forms which fall within its scope will differ to a greater or lesser extent in their amino acid sequences. In turn, these differences in primary structure will also entail greater or lesser differences in the higher levels of protein structure. The interpretation of the differences between isoenzymes in structural terms is well advanced in several cases. However, the origins of other categories of enzyme heterogeneity, and therefore the differences in structure existing within them, are in general much less clearly understood.

ORIGINS OF ISOENZYMES

The groups of genes which determine the structures of families of isoenzymes can represent several different phenomena: the existence of multiple gene loci, the occurrence as the result of mutation of pairs of unlike genes (alleles) at the same locus, or the modification of the structures or expression of genes in somatic cells, e.g. as an accompaniment of malignant transformation (Fig. 2.1).

Isoenzymes which are the products of allelic genes are distributed in the population according to the laws of Mendelian inheritance, and these hereditary patterns identify the nature of their genetic origins. Multiple forms of enzymes resulting from the existence of multiple gene loci have become disseminated throughout the whole species during the course of evolution, so that all individuals typically possess the same complement of isoenzymes. Consequently the genetic origins, and therefore the iso-enzymic status of the multiple forms, cannot be readily inferred by comparing their patterns of occurrence. In some cases however, allelic

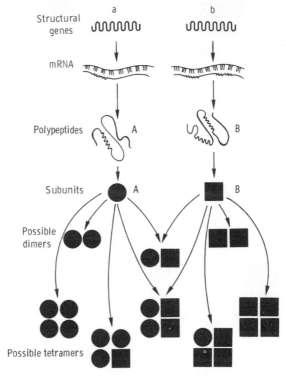

Fig. 2.1   *Origin of isoenzymes. Two or more genes (a and b) determine the structure of distinct polypeptides A and B. The polypeptides may themselves constitute monomeric isoenzymes, or they may be the subunits of polymeric isoenzymes. When the structures of the polypeptides are sufficiently similar, further hybrid isoenzymes may be added to the respective homopolymers (From Moss, 1979. By permission of the Chemical Society, London).*

variation at a particular locus may have conferred recognizable, inherited modifications on its products, thus indicating their related genetic origins and distinguishing them from the analogous products of other loci. For example, the form of alkaline phosphatase which occurs in the human placenta exhibits numerous allelic variants (Donald and Robson, 1974). Since this variation is not reflected in the alkaline phosphatases of other tissues, the structure of the placental isoenzyme, at least, must be controlled by a separate gene locus. Mutation at some loci seems to be particularly rare, especially in man, and such evidence of the genetic basis of multiple enzyme forms is correspondingly scanty. Nevertheless, confirmation that some isoenzymes found in human tissues are the products of separate gene loci has been obtained in this way. The products of each of the main human lactate dehydrogenase loci (i.e. those determining the H- and M-subunits) have been found to be independently modified in a few individuals (Boyer *et al.*, 1963; Nance *et al.*, 1963). Similarly, rare variants of the mitochondrial isoenzyme of human aspartate aminotransferase can occur without corresponding changes in the cytoplasmic isoenzyme, and are inherited in a Mendelian manner (Davidson *et al.*, 1970).

Evidence for the genetic origins of multiple forms of enzymes can be obtained in some cases from the artificial transfer of genetic material in cell hybridization experiments. Often, however, evidence of the natural or artificial inheritance of differences between enzyme forms is lacking, so that it becomes necessary to compare the structures and properties of multiple enzyme forms in attempting to determine whether they originate from genetic heterogeneity or by post-genetic modification. These approaches are discussed more extensively in later sections.

Isoenzymes determined by multiple gene loci

A substantial proportion of enzymes are determined by several structurally different gene loci, and therefore exist in isoenzymic forms. A survey of evidence relating to 66 human enzymes showed no fewer than 24 to be the products of more than one gene locus, with three loci being involved in determining the structures of nine enzymes (Hopkinson *et al.*, 1976). It is possible that the structures of human alcohol dehydrogenase and hexokinase are each coded by as many as four gene loci. Among the numerous examples of isoenzymes determined by multiple gene loci are those forms of several enzymes which are characterized by their specific intracellular locations, such as the cytoplasmic and mitochondrial forms of aspartate aminotransferase or NAD-dependent malate dehydrogenase, as well as isoenzyme systems with similar intracellular localizations but with more or less tissue-specific distributions; e.g. the isoenzymes of lactate dehydrogenase, which are determined by three loci in human tissues; aldolase, also the product of three loci; creatine kinase, determined by two loci, and many more.

Multiple gene loci which determine the structures of functionally-similar enzymes may have come into existence as a result of gene duplication during the course of evolution, followed by independent mutation of the separate loci. Alternatively, originally distinct genes may have converged through successive mutations, with the result that their dependent enzymes have developed similar catalytic functions. The various classes of proteases, those containing a serine residue at the active centre, those with an active sulphydryl group, and the metal-dependent proteases, may have originated in this way.

A greater degree of functional and structural similarity might be expected in the case of isoenzymes originating by divergent evolution following gene duplication, than for the 're-invention' of a particular catalytic process which convergent evolution may represent; indeed, the various classes of proteases which appear to have resulted from the latter type of evolutionary process are not usually considered to be isoenzymes. However, only tentative inferences can be drawn as to the possible evolutionary origins of the genes determining functionally similar enzymes solely on the basis of a comparison of the catalytic specificities of enzyme variants.

The occurrence of genes determining a group of isoenzymes or functionally-similar enzymes in close proximity to each other on the chromosomes is evidence that they have probably originated by gene duplication, especially when the linkage is evident in several species. The two loci determining isoenzymes of α-amylase are closely linked in man

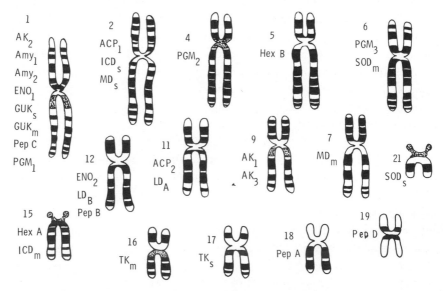

Fig. 2.2   Assignment of multiple loci determining individual isoenzymes to various human chromosomes. $ACP_{1,2}$, acid phosphatase; $AK_{1,2,3}$, adenylate kinases; $Amy_{1,2}$, salivary and pancreatic amylases; $ENO_{1,2}$, enolases; $GUK_{1,2}$, soluble and mitochondrial guanylate kinases; Hex A,B, hexosaminidases; $ICD_{s,m}$, soluble and mitochondrial isocitrate dehydrogenases; LD $_{A,B}$, lactate dehydrogenases; $MD_{s,m}$, soluble and mitochondrial maltate dehydrogenases; Pep A,B,C,D, peptidases; $PGM_{1,2,3}$, phosphoglucomutases; $SOD_{s,m}$, soluble and mitochondrial superoxide dismutases; $TK_{s,m}$, soluble and mitochondrial thymidine kinases.

and in other species, as are carbonic anhydrase loci in several non-human vertebrates. Groups of genes determining non-specific esterases are linked in the mouse and rat (Shows, 1977). However, in other instances in which gene duplication appears to be the most probable explanation for the existence of isoenzyme-determining genes, the genes are not in close proximity to each other and may even be carried on different chromosomes. This is true, among other examples, of the genes determining the structures of the H- and M-subunits of human lactate dehydrogenase, which are respectively located on chromosomes 12 and 11 (Fig. 2.2).

Isoenzymes determined by multiple alleles

A large number of enzymes in many species exist in multiple molecular forms which differ in characteristics or distribution from one individual to another. The multiple forms originate from the existence of modified genes, or alleles, at various chromosomal loci. The isoenzymes determined by allelic genes have therefore also been termed 'allelozymes' (or 'allozymes'). Family studies show that these individual isoenzyme patterns are inherited according to Mendelian laws.

The alleles determining certain variant isoenzymes may occur with frequencies which are appreciable when compared with that of the most usual variant. The population is then said to be polymorphic with respect to the isoenzymes. Other allelozymes may occur only with extreme rarity.

In contrast to inheritance of characters such as eye colour, the isoenzyme (or its component subunit) produced by one allele does not dominate or mask the product of an unlike allele with which it may be paired; thus, whereas individuals who are homozygous at a particular locus exhibit only the gene product characteristic of a single allele, the products of both the allelic genes are expressed in heterozygous individuals.

The proportion of enzyme-determining loci which are subject to allelic variation has been estimated to be as high as 28% based on differences in the electrophoretic mobilities of enzymes determined by 71 human gene loci (Harris and Hopkinson, 1972). This is probably a conservative estimate in view of the relatively few loci, species and populations for which this kind of systematic survey has been made. Furthermore, not all the modifications which an enzyme molecule may undergo will be detectable by electrophoresis. Within a particular species, allelic variation seems to be more common at some loci than others. The prevalence within the population of individuals heterozygous at a particular locus can be taken as a measure of the mutation rate at that locus. The range of heterozygosities for enzyme variants within species is wide – from zero for some enzymes to about 0.80 for others. There is a tendency for enzymes which are structurally invariate in one species to be similarly devoid of variation in other species. Several factors may contribute to this non-uniform distribution of mutations among enzyme-determining genes.

It would seem reasonable to infer that changes in structure are more likely to occur in enzymes with larger rather than smaller polypeptide chains, since the opportunities for mutations which modify, but do not abolish, the character of the product should be greater in the case of larger structural units. A positive correlation between heterozygosity at enzyme-determining loci and the polypeptide molecular weights of their dependent isoenzymes has been demonstrated for Drosophila species and for Heliconiine butterflies, and some data in support of the hypothesis has also been obtained for certain vertebrate species (Koehn and Eanes, 1979). However, other studies have failed to find the expected relationship between heterozygosity and polypeptide molecular weight in *Drosophila* or *Colias* butterfly (Johnson, 1977), and the evidence from studies of human enzyme polymorphism is similarly equivocal. Harris *et al.* (1977) found no significant difference between the mean polypeptide sizes of non-variate and polymorphic human enzymes. On the other hand, Koehn and Eanes (1979) found a significant correlation between the number of rare alleles and the subunit sizes of the enzymes concerned.

The failure to find as clear a relationship as might be expected between polypeptide size and number of enzyme variants suggests the existence of determinants other than the size of the gene. Such determinants may take the form of selection pressures opposing the persistence of certain modifications, or they may have a basis in isoenzyme structure. Many if not most enzymes are composed of several polypetide subunits; often two or four, and more rarely three. The regions of the polypeptides involved in contacts between subunits are highly conserved (Klotz *et al.*, 1970) and the formation of hybrid isoenzymes between homologous subunits from different species discussed later is evidence for this. About 14% of the surface area of each monomer is involved in dimer formation, and correspondingly more in tetrameric molecules (Teller, 1976). Since the active form of the enzyme is almost invariably the oligomer, mutations which prevent subunit association will abolish catalytic activity. The contact areas therefore represent a considerable part of the polypeptide structures in which the possibilities of structural alterations must be greatly limited. Evidence that this is so has been found for allelozymes of human enzymes, in that much lower average heterozygosities are seen at a series of loci determining oligomeric isoenzymes than at loci whose products are monomeric, although the average size of polypeptides in the two classes is not greatly different (Harris *et al.*, 1977).

Hybrid isoenzymes

Active molecules of oligomeric enzymes may arise by the association of similar but non-identical subunits, either *in vivo* or *in vitro*. When the different subunits are the products of separate structural genes, the hybrid molecules thus formed are themselves included in the formal definition of isoenzymes. The different subunits concerned may be the products of separate gene loci or of allelic genes at the same locus (Fig. 2.1). Artificial hybrid 'isoenzymes' may be produced by re-combination experiments involving modified subunits.

The number of different hybrid isoenzymes which can be formed from two non-identical protomers depends on the number of subunits in the complete enzyme molecule. If the number of different subunits is *s* and the isoenzyme molecules are each composed of *n* subunits, the number of different isoenzymes which can be formed, assuming that all combinations of subunits are possible, is given by $(s+n-1)!/n!(s-1)!$. Thus, for a dimeric enzyme, one mixed dimer may be added to the two dimers composed of pairs of identical subunits if two different subunits exist, while for an enzyme with four subunits the formation of three heteropolymeric isoenzymes is possible from two types of subunits. Among the many well-studied examples of hybrid isoenzymes are the mixed MB dimer of human creatine kinase, consisting as its designation indicates, of one M and one B subunit, and the three hybrid isoenzymes LD_2, LD_3 and LD_4 of lactate dehydrogenase. The latter have the subunit compositions H_3M, H_2M_2 and HM_3, respectively, in which H and M (or alternatively B and A) represent protomers produced by separate gene loci. The subunit (X or C) determined by the third lactate dehydrogenase locus in man and some animals can also enter into hybrid isoenzyme formation *in vitro* and *in vivo*, though this does not happen in human tissues.

The generation of hybrid isoenzymes accounts for part of the considerable complexity of isoenzyme zones which may be encountered in the electrophoretic analysis of polymorphic enzymes in tissue extracts. When the isoenzymes produced by allelic genes are monomeric (or, in the case of multimeric enzymes, are so unalike that hybrid multimers cannot be

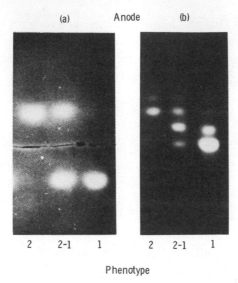

Fig. 2.3(a)   *Variants produced by allelic genes at the locus which determines isoenzyme II of human carbonic anhydrase. Only a single zone is seen in each of the extracts of erythrocytes from individuals of phenotypes 1 and 2, who are homozygous at this locus. Both zones are present in the heterozygote (phenotype 2–1). Separation was by starch-gel electrophoresis. The substrate was fluorescein diacetate, which is preferentially hydrolysed by isoenzyme II (From Hopkinson et al., 1974. By permission of Cambridge University Press).*
Fig. 2.3(b)   *Formation of a hybrid dimeric isoenzyme of esterase D in individuals of phenotype 2–1 who are heterozygous at the locus determining this enzyme. Homozygotes (phenotypes 1 or 2) each exhibit one main allelozyme with secondary zones which probably result from post-translational modification. The zones were separated by starch-gel electrophoresis and treated with the fluorigenic substrate 4-methyl umbelliferyl acetate which is preferentially hydrolysed by esterase D (From Hopkinson et al. (1973). By permission of Cambridge University Press).*

formed), the isoenzyme patterns in the tissues of heterozygous individuals consist of the sum of the patterns characteristic of the respective homozygotes (Fig. 2.3). When, however, hybridization between the products of allelic genes can occur, one, two or more hybrid isoenzymes are added to the respective homopolymeric isoenzymes, depending on the number of subunits in the complete enzyme molecules (Fig. 2.3). The patterns can be further complicated when a particular enzyme activity is determined by multiple gene loci. If allelic variation occurs at one or more of these loci, additional possibilities of hybrid isoenzyme formation may arise.

STRUCTURAL DIFFERENCES BETWEEN ISOENZYMES AND THEIR INVESTIGATION

As already mentioned, the inclusion of the multiple forms of a particular enzyme within the category of isoenzymes implies that they differ in primary structure to some degree, however slight, and consequently that differences will probably also exist at the higher levels of protein structure. Full elucidation of these structural differences requires the application of the chemical and physical methods of protein analysis to purified preparations of the individual isoenzymes. Although considerable progress has been made in the case of a few families of isoenzymes, the rate at which such definitive information can be gathered is restricted in most cases by the limited amounts and incomplete purities of isoenzyme samples, particularly of those which occur in human tissues or which are due to rare allelic genes. Analytical methods which can be applied to small, impure isoenzyme samples, but from which structural inferences can nevertheless be drawn, are therefore particularly valuable.

Differences in primary structure

Comparative amino acid sequences of isoenzymes have so far been determined in a few cases only. The primary structures of the B and C forms of human carbonic anhydrase (isoenzymes I and II) differ by about 90 amino acid residues out of the 260 and 259 which respectively make up the single polypeptide chains of these isoenzyme molecules (Andersson *et al.*, 1972; Giraud *et al.*, 1974; Henderson *et al.*, 1973), with more than half of the sequences of two or more residues being identical. The longer M- and H- chains of lactate dehydrogenase from porcine tissues (331 and 333 residues respectively) similarly show considerable homologies of primary structure (Kiltz *et al.*, 1977). The cytoplasmic and mitochondrial isoenzymes of aspartate aminotransferase from pig heart are each dimeric molecules with subunits consisting of 412 and 401 amino acid residues respectively. Nearly half of their primary structures are homologous, with 42 sequences of two or more residues in common (Doonan *et al.*, 1974; Kagamiyama *et al.*, 1977).

Peptide maps and partial sequences

Partial information about the primary structures of isoenzyme molecules can be obtained by the identification of carboxy- or amino-terminal amino acid residues and possibly of short sequences of amino acids adjacent to these residues, when purified preparations are available. Identical N-terminal sequences, extending to four residues, were found for normal human-placental alkaline phosphatase and a tumour-derived variant, but these differed from the corresponding sequence for liver alkaline phosphatase (Greene and Sussman, 1973; Badger and Sussman, 1976). Stepwise removal of N-terminal amino acids from component peptides of enzyme molecules, e.g. by the Edman degradation, can usually now be repeated for about 10–20 residues with the aid of techniques in which the peptide is attached to a solid phase. However, the most sensitive and specific indication of differences in primary structures of closely similar proteins such as isoenzymes, short of determination of complete amino acid sequences, is given by comparison of two-dimensional maps of peptides obtained by partial hydrolysis of the proteins with enzymes or acid. This 'fingerprinting' technique is capable of revealing the presence of the single amino acid substitutions which are, in many cases, the only differences in primary structure between the products of allelic genes. The fingerprint technique can be used to identify peptides in which differences in amino acid sequences between isoenzymes occur, so that sequence determination can be concentrated on these peptides. In this way, single amino acid substitutions have been shown to account for the differences between allelozymes of human glucose-6-phosphate de-hydrogenase (Yoshida, 1967) and between usual and variant forms of carbonic anhydrases B (Tashian *et al.*, 1966; Funakoshi and Deutsch, 1970) and C (Lin and Deutsch, 1972). Other examples of the application of finger-printing to isoenzyme analysis include studies of the isoenzymes of lactate dehydrogenase (Wieland *et al.*, 1964; Chang *et al.*, 1979), alkaline phosphatase (Badger and Sussman, 1976) and creatine kinase (Dawson *et al.*, 1968), and comparison of the normal form of the latter enzyme with a variant present in muscle of dystrophic mice (Hooton and Watts, 1966). Identical maps were obtained from the multiple forms of mitochondrial aspartate aminotransferase (Michuda and Martinez-Carrion, 1969).

Cleavage with enzymes such as trypsin or pepsin is preferable to partial hydrolysis with acid or alkali because the specificity of proteolytic enzymes ensures that their action is reproducible and non-random. However, long periods of digestion with enzymes or even an initial partial denaturation by heat, acid, or other agents may be required, since some isoenzyme proteins are markedly resistant to proteolysis. Separation of peptides from undegraded protein by gel filtration is a useful preliminary to separation by high-voltage electrophoresis or chromatography, or both. Plates coated with thin layers of alumina or cellulose for two dimensional separations, or polyacrylamide gels for unidimensional electrophoresis, have now generally replaced the filter-paper of earlier studies as supporting media for resolution of the peptide mixtures.

Comparison of maps stained with detection reagents such as ninhydrin or fluorescamine assumes that all the peptides are indeed derived from the isoenzymes being analysed, i.e., that each isoenzyme is free of other proteins, a condition that is difficult to satisfy when abundant sources of the isoenzymes are not available. However, the specificity of the method can be improved by attaching a label to the isoenzyme molecules before partial hydrolysis, the presence of which subsequently identifies those peptides containing it as having been derived from the isoenzyme and not from protein impurities. Several labels have been selected for this purpose on the basis of their affinities for specific sites, usually the active centres of the isoenzyme molecules. The presence of pyridoxal phosphate as a prosthetic group at the active centre of aminotransferases provides a means of identifying peptides derived from this molecular region. Radioactive peptides containing pyridoxal phosphate linked to lysine have been obtained by reducing bovine mitochondrial aspartate aminotransferase with tritiated lithium borohydride, then digesting the isoenzyme with the proteolytic enzyme thermolysin (Bossa *et al.*, 1976). The amino acid sequence of the radioactive peptide was identical to that of peptides from corresponding isoenzymes from the mitochondria of other species. Active-centre peptides from cytoplasmic aspartate aminotransferase show inter-species identity, but with some differences between the cytoplasmic and mitochondrial isoenzymes. At acid pH, alkaline phosphatases incorporate orthophosphate rapidly and irreversibly, with formation of phosphorylserine. The covalent bond thus formed survives partial hydrolysis of the enzyme protein by proteolytic enzymes or acid. Radioactive peptides have been isolated from alkaline phosphatases from micro-organisms and from human and animal tissues after labelling the enzymes with ^{32}P-orthophosphate (Schwartz *et al.*, 1963; Milstein, 1964; Engstrom, 1964; McKenna *et al.*, 1979; Whitaker and Moss, 1979).

It is to be expected that the structures of the regions immediately surrounding specific ligand-binding sites of enzyme molecules will have been strongly conserved during the evolutionary processes which have led to the emergence of isoenzymes. This indicates a possible limitation on the comparison of peptides derived from these regions: unless the different peptides are large enough to encompass portions of the primary structure which are relatively remote from the active site, it is perhaps

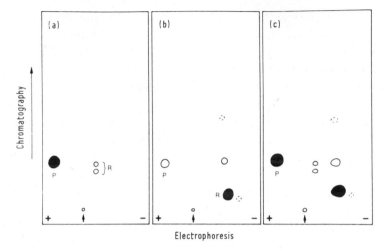

Fig. 2.4   *Tracings of autoradiograms of two-dimensional separations of tryptic digests of alkaline phosphatases from (a) human kidney and (b) human placenta after labelling with* ^{32}P-*orthophosphate. Different radioactive peptides (R) are obtained from the two isoenzymes. P is unbound phosphate; (c) is a mixture of the two digests (From Whitaker and Moss, 1979. By permission of the Biochemical Society).*

unlikely that differences in amino acid sequence will be found. However, true isoenzymes (in contrast to multiple forms arising by post-genetic modification) typically possess slightly different catalytic properties (Chapter 3) which presumably reflect minor structural differences at or near the binding sites. Radioactive peptides separated after tryptic digestion of human kidney alkaline phosphatase labelled with ^{32}P-orthophosphate are different from the single peptide obtained in a similar way from the genetically-distinct placental isoenzyme (Whitaker and Moss, 1979) (Fig. 2.4). Isoenzymes produced by allelic mutation also often possess altered catalytic characteristics, and different patterns of radio-active peptides were obtained by electrophoresis of partial hydrolysates of the usual and atypical (dibucaine-resistant) allelozymes of human serum cholinesterase, after these isoenzymes had been labelled with an active-centre-directed reagent, radioactive di-isopropyl fluorophos-phonate (Muensch et al., 1978).

Selective chemical or enzymic modification

Many of the ligands useful in labelling active-site peptides can also be employed to identify particular groups which form part of the active centres of isoenzymes. However, since differences between isoenzymes in the characteristics of their active centres are generally quantitative rather than qualitative, active-centre-specific ligands find their place in isoenzyme analysis mainly in measurements of relative affinities of isoenzymes for them, rather than in the exploration of the topography of binding sites. Several chemical or enzymic modification procedures can provide evidence of the presence in isoenzyme molecules of particular chemical groupings or linkages in regions other than the active centre. In some cases, the altered molecules thus produced resemble the multiple forms of an enzyme found in extracts of cells or tissues, suggesting ways in which post-genetic modifications may occur *in vivo*.

The most useful modifications of isoenzyme molecules are those which result in some marked change in properties while preserving catalytic activity, since this permits the study of the effects of modification even when only impure or limited isoenzyme samples are available. Methods which produce changes in net molecular charge are particularly advantageous because of the resulting alterations in the electrophoretic mobilities of catalytically-active isoenzyme zones. The net charge of protein molecules is determined by the numbers and states of ionization of several types of amino acid side chains in contact with the aqueous environment, principally the amino groups of lysine, hydroxylysine and arginine and the carboxyl groups of aspartic and glutamic acids, with, to a lesser extent, the hydroxyl groups of serine, threonine and tyrosine, the imidazole ring of histidine, and the sulphydryl group of cysteine. The ionization of these groups can be modified or prevented by chemical treatments of varying degrees of specificity, so that the selective effects of such treatments on the electrophoretic mobilities of individual isoenzymes can be interpreted in terms of the probable relative numbers of the modifiable residues present in their respective molecules. Examples of modification procedures include esterification of carboxyl groups, acylation of amino groups with acetic or succinic anhydrides, nitration of tyrosine residues with tetranitromethane, and reaction of sulphydryl groups with reagents such as *N*-ethyl maleimide. Some loss of enzymic activity may result from non-specific denaturation during these treatments, although the necessary conditions are usually mild. With some enzymes, groups essential for substrate binding or conversion may be modified, with loss of activity, although this may be prevented to some extent by carrying out the reaction in the presence of the substrate or a competitive inhibitor.

An example of a differential effect of chemical modification on the electrophoretic mobility of isoenzymes is the observation that acetylation of alkaline phosphatases from human placenta and small intestine increases their anodal mobility to a greater extent than is the case for hepatic phosphatase similarly modified (Moss, 1970a). Since acetylation under the chosen conditions probably selectively modifies the amino groups of lysine and arginine, it may be inferred that these groups make a smaller contribution to the net charge of native hepatic alkaline phosphatase than is the case for the other two isoenzymes. Carbamoylation of amino groups with sodium cyanate at pH 9.7 and 37°C produces similar differential effects. Acetylation or succinylation of the three most common allelic variants of placental alkaline phosphatase, produces a greater acceleration of the slow (S) isoenzyme than of either the fast (F) or intermediate (I) isoenzymes (Thomas and Moss, 1972).

Modification of the properties of isoenzymes by treatment with enzymes offers the advantage of high specificity which is inherent in other uses of enzymes as analytical reagents. Human pancreatic amylase exhibits a series of zones on electrophoresis, the least anodal fraction being the most prominent. Deamidation, with a resultant increase in the number of ionizable carboxyl groups and thus of the net negative charge, has been suggested as the explanation of the more anodal bands (Karn et al., 1974). Experimental confirmation of this hypothesis has been obtained by enzymic modification of purified human pancreatic amylase with bacterial peptidoglutaminases (Ogawa et al., 1978). The two peptidoglutaminases used in this work, from B. circulans, have somewhat different specificities: although both convert glutamine residues in polypeptides to glutamic acid, one acts on residues closer to the carboxy-terminus of chains than the other. Both produced modified pancreatic amylase zones similar to those seen when pancreatic juice itself is incubated, but at rather different rates, suggesting the existence of several modifiable glutamine residues located at different sites in the amylase molecule.

Secondary and tertiary structures

These levels of structure of protein molecules are themselves apparently determined by the linear sequence of amino acids in a particular polypeptide chain, i.e. by its primary structure, since the chain tends to assume its most stable configuration. Therefore, differences between isoenzymes in secondary and tertiary structures are to be expected, although considerable similarities will exist when variations between the respective amino acid sequences are few, or when one amino acid residue is replaced by another of similar configuration and characteristics. The elucidation of the three-dimensional structures of isoenzyme molecules, as of other proteins, can be achieved by X-ray crystallography. Since this definitive technique requires the preparation of crystals of isomorphous heavy-atom-substituted forms, as well as of the native molecules, its applications to the comparison of isoenzyme structures are still few in number. Those studies which have been made show the general similarities between isoenzymes which would be anticipated from the degree of homology of their primary structures. Nevertheless, detailed differences in spatial arrangement do exist, and these can provide important insights into the structural and functional relationships between isoenzymes.

The cytoplasmic and mitochondrial aspartate aminotransferase iso-enzymes from chicken tissues studied at 4–5 Å resolution possess an overall similarity in mode of arrangement of the subunits, but are not identical in structure (Borisov *et al.*, 1978; Eichele *et al.*, 1979). A comparison of the three-dimensional structures of human erythrocyte carbonic anhydrase isoenzymes B and C at 2 Å resolution shows that these molecules also have the overall conformational similarity which would be expected from the considerable homologies between their primary structures (Notstrand *et al.*, 1975). Thus, there are 10 bends in the polypeptide chain occurring at the same points in the two molecules, and the large elements of β-structure are identically located. There are few amino acid differences in the hydrophobic cores of the molecules.

The active site cavity of the carbonic anhydrases, containing the zinc atom, consists of hydrophobic and hydrophilic halves, and there are a number of differences in amino acid residues in the respective hydro-phobic regions of the cavities of the two isoenzymes as well as some differences in the hydrophilic parts. The effect of the differences is to reduce the available volume of the cavity in the neighbourhood of the zinc atom in isoenzyme B, compared with isoenzyme C, so that they probably underlie the different catalytic properties of the two isoenzymes.

As is the case for the different isoenzymes of carbonic anhydrase, the lactate dehydrogenase isoenzyme M_4 from dogfish and isoenzyme H_4 from pig heart resemble each other closely in overall conformation, as revealed by X-ray diffraction analysis of their ternary enzyme–coenzyme–substrate analogue complexes, but again, significant differences occur in critical substrate-binding regions. An alanine residue at the active centre of the muscle isoenzyme is replaced by glutamine in the isoenzyme from heart, affording a further opportunity for hydrogen bond formation to the nicotinamide phosphate in the latter case. Changes in the H_4 isoenzyme increase the volume of the adenosine-binding pocket compared with M_4 lactate dehydrogenase (Eventoff *et al.*, 1975; 1977).

When samples of purified isoenzymes are available, the presence of certain three-dimensional structures such as α-helices can be deduced with the aid of physical techniques of more limited scope than X-ray diffraction analysis. Measurement of optical rotatory dispersion for three chromatographically-separated forms of mitochondrial aspartate aminotransferase from pig heart indicated that these forms have similar conformations (Michuda and Martinez-Carrion, 1969). This method was also used to investigate postulated differences in conformation amongst the sub-forms of malate dehydrogenase from chicken mitochondria which can be separated electrophoretically, and to compare the native and iodinated forms (Kitto *et al.*, 1966). The results suggested that, of the naturally-occurring forms, the one which is most cathodal on electrophoresis possessess the greatest content of helical structures. The introduction of between approximately one and three atoms per mole of iodine into this molecular species produced differences in optical rotatory dispersion comparable to those observed amongst the less cathodal native forms, possibly due to partial unfolding of secondary and tertiary structures. Iodination also resulted in the appearance of forms with electrophoretic mobilities similar to those of the native variants.

Limited comparisons of localized conformational differences between isoenzymes can also be made by the use of specific ligands which become attached to specialized regions of enzyme molecules, offering a means of exploring the topographical differences between analogous regions of isoenzymic molecules. Differences in the properties of binding sites for substrates, cofactors or inhibitors underlie many of the variations between isoenzymes in their catalytic behaviour discussed in the next chapter and, although some of these inter-isoenzymic variations are due to the presence of different chemical groupings, others must derive from differences in the three-dimensional structures of active sites. Thus, the exploration of ligand-binding sites involves consideration of secondary and tertiary structures, as well as primary structure.

A particular structural feature is common to enzymes which require nucleoside phosphates as coenzymes or substrates. This is the 'dinucleotide fold', a zone of secondary and tertiary structure forming non-polar pockets into which the aromatic rings of the nucleotides fit. The dye Cibachron Blue F3GA has structural resemblances to nucleotides and can therefore become bound to the dinucleotide fold of enzymes and isoenzymes which possess this feature. Attachment of the dye is accompanied by a spectral change which can be detected by difference spectrophotometry. Dissimilar dissociation constants for the binding of Cibachron Blue by the M_4 and H_4 isoenzymes of lactate dehydrogenase, from rabbit muscle and beef heart respectively, have been reported (Thompson and Stellwagen, 1976), suggesting that the topography of the dinucleotide fold is not identical in these two isoenzymes.

Maintenance of the three-dimensional structures of protein molecules depends mainly on the large numbers of hydrogen bonds and hydrophobic interactions which exist between adjacent amino acid side chains, together with covalent linkages such as disulphide bridges in some cases. Differences between isoenzymes in their resistance to denaturing agents (Chapter 3) reflect variations in the number and strength of these stabilizing interactions, thus providing what is probably the most readily-obtainable evidence of conformational differences. However, such observations cannot be interpreted in terms of specific structural differences.

Conformational isomerism

The observations on the effects of iodination on the multiple forms of mitochondrial malate dehydrogenase gave rise to the hypothesis that conformational isomerism might constitute a general cause of enzyme heterogeneity, since, at least in theory, several different configurations of a single polypeptide chain may have nearly equal stabilities (Kitto *et al.*, 1966). Minor forms of other enzymes or isoenzymes, e.g., of aspartate aminotransferase, the MM-isoenzyme of creatine kinase, and erythrocyte acid phosphatase, have also been interpreted as conformational isomers, or 'conformers'. Modifications in electrophoretic mobility produced by treatment with denaturing agents or by separation in the presence of ligands have similarly been ascribed to conformational changes. Examples of the latter effect include the altered mobilities of active zones of alcohol dehydrogenase from human or horse liver or from *Drosophila melanogaster* in the presence of the coenzyme, nicotinamide adenine dinucleotide (Smith *et al.*, 1971a; McKinley-McKee and Moss, 1965; Jacobson, 1968) but, as with other postulated conformational isomerisms, alternative explanations can be put forward.

A prediction of the conformer hypothesis is that reversible denaturation of one of the isomers should generate the complete set, since it is implied that the several alternative conformations are of nearly equal stabilities. However, attempts to test this prediction have not been completely successful (Schechter and Epstein, 1968). Although reversible denaturation of individual mitochondrial isoenzymes of malate dehydrogenase with acid or guanidine hydrochloride did produce renatured forms with electrophoretic mobilities similar to those of native isoenzymes, the native and renatured forms were not identical in stability to heat (Kitto *et al.*, 1970). In no case can the existence of conformers be regarded as unequivocally established, to the exclusion of other possible explanations of a multiplicity of stable, coexisting enzyme forms.

Quaternary structure

Probably the majority of enzyme molecules consist of aggregates of smaller subunits or monomers, each consisting of a single polypeptide chain; i.e., the molecules possess a quaternary level of structure.

Association of non-identical subunits in various combinations to form catalytically-active molecules frequently accounts for the existence of multiple forms of enzymes. As already mentioned, when the unlike subunits are themselves the products of different genes, either allelic or multi-locular, the hybrid molecules are considered to be isoenzymes, as are the homopolymers of the respective subunits. The formation of hybrid isoenzymes *in vitro* by the association of protomers originating in different species is evidence of conservation of the structures of the contact surfaces of the protomers, as it is in the case of naturally-occurring hybrid isoenzymes formed between the polypeptides produced by multiple-gene loci or alleles.

Methods for investigating the quaternary structure of enzyme molecules occupy an important place in the study of multiple forms of enzymes. Dissociation of polymeric protein molecules into their component monomers with the aid of reagents such as urea or guanidine hydrochloride, with in some cases the additional use of reagents which break disulphide bridges, is an established procedure in the structural analysis of proteins, e.g., in determining the sizes of constituent subunits or as a preliminary to sequence analysis. However, the subunits of isoenzyme molecules thus produced are usually catalytically inactive, so that the parent isoenzymes must be available in quantities and states of purity which allow the detection and identification of the dissociated subunits by methods independent of enzymic activity. These limitations are avoided when dissociation of one or more types of isoenzyme molecules is followed by a reversal of the conditions, so as to permit reassociation of the protomers into new polymeric combinations, each with catalytic activity. From the number and properties of these the probable quaternary structures of the original polymers can be inferred. Hybridization experiments of this kind can be carried out in various ways:

Hybridization in vitro

When a mixture of isoenzymes 1 and 5 of lactate dehydrogenase in equal proportions is frozen in 1 mol l^{-1} sodium chloride solution and then thawed, the two homopolymers dissociate into their component H- and M-subunits which re-combine to produce the five possible tetramers (H_4, H_3M, H_2M_2, HM_3 and M_4) in the proportions expected for random reassociation. The appearance of the hybrid isoenzymes can be demonstrated by zone electrophoresis (Markert, 1963). This type of experiment has been widely applied and extended, not only to the production of inter- and intra-species hybrid molecules of lactate dehydrogenase (Salthe *et al.*, 1965; Zinkham *et al.*, 1963), but also to many other isoenzyme systems (Fig. 2.5). Although mixtures of the homopoly-

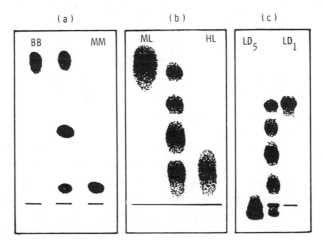

Fig. 2.5 *Diagrams of experimental formation of hybrid isoenzymes. The mixtures were separated by starch gel electrophoresis (anode at the top), with the hybridized sample in the centre.* (a) *Rabbit muscle (MM) and brain (BB) creatine kinase isoenzymes, showing the formation of one additional mixed (MB) dimer, based on data from Dawson et al. (1967).* (b) *Mouse (ML) and human (HL) liver nucleoside phosphorylases, showing formation of two hybrid isoenzymes of this trimeric enzyme, based on data of Edwards et al. (1971).* (c) *Ox heart LD₁ and LD₅ isoenzymes of lactate dehydrogenase, showing formation of three mixed tetramers, from an experiment by Markert (1963) (From Moss, 1979. By permission of the Chemical Society, London).*

meric isoenzymes are often used as sources of the different subunits, dissociation of a single type of heteropolymer with subsequent generation of new isoenzymic combinations has also been achieved. Tetrameric structures have been inferred for aldolase (Penhoet *et al.*, 1966), malic enzyme (Li, 1972), phosphofructokinase (Tsai and Kemp, 1972), pyruvate kinase (Cardenas and Dyson, 1973) and hexosaminidase (Beutler and Kuhl, 1975; Beutler *et al.*, 1976) by *in vitro* hybridization. Isoenzymes shown in this way to be dimers include liver alcohol dehydrogenase (Lutstorf and von Wartburg, 1969), glucose-6-phosphate dehydrogenase (Yoshida *et al.*, 1967) and creatine kinase (Dawson *et al.*, 1965), and a trimeric structure has been demonstrated for nucleoside phosphorylase (Edwards *et al.*, 1971).

Although mild conditions such as freezing and thawing in solutions of high ionic strength induce dissociation and recombination in many cases, more vigorous treatment may sometimes be necessary; e.g. inter- and intra-species hybridization of bovine and chicken pyruvate kinases required, first, dissociation with guanidine hydrochloride, followed by renaturation by dilution into buffer with dithiothreitol or β-mercaptoethanol (Cardenas *et al.*, 1975), while treatment with 8 mol l^{-1} urea solution for 30 min with subsequent removal of urea by dialysis in the presence of β-mercaptoethanol was needed to produce hybrid creatine kinase molecules containing two forms of the M-subunit (Wevers *et al.*, 1977). Dissociation and recombination during electrophoresis has been effected in the case of isoenzymes of the cytoplasmic form of superoxide dismutase by carrying out the electrophoretic separation in starch gel at a temperature of 40°C (Edwards *et al.*, 1978). At lower temperatures, extracts of liver tissue from subjects heterozygous for the enzyme show the three-banded pattern expected for a dimeric enzyme such as this. At the higher temperature, however, only zones corresponding to the homopolymers are seen. This is explained by the existence of an equilibrium between the dimeric isoenzymes and their constituent monomers above about 35°C. Since the two types of monomer are undergoing separation due to their different net charges during electrophoresis, reformation of the homodimers can occur, but not of mixed dimers.

In these applications of the *in vitro* hybridization technique, the different subunits which are induced to dissociate and recombine are the naturally-occurring products of separate gene loci or alleles. However, in one case, in which no natural isoenzyme exists with electrophoretic properties distinct from those of the isoenzyme under study with which hybridization experiments can be carried out, enzymic modification has been used to produce a set of subunits with altered net charge. This study concerned the C_4 component of cholinesterase, which constitutes the major zone of this enzyme seen after electrophoresis of serum in starch gel (Scott and Powers, 1972). The isoenzyme purified from human plasma was treated with neuraminidase to reduce its net negative charge at alkaline pH by removal of terminal N-acetyl neuraminic acid residues. Hybridization of the modified isoenzyme with its native counterpart was achieved by freezing and thawing three times, a mixture of the two types of molecules in a solution of sodium chloride ($4 \, mol^{-1}$) and mercaptoethanol (1% w/v) at pH 10–11, then keeping the mixture at $4°C$ for at least 24 h. Subsequent electrophoresis showed the unmodified and modified isoenzymes together with three new components migrating between them, a result consistent with a tetrameric structure of the native cholinesterase isoenzyme.

Hybridization in living cells

Hybrid isoenzymes occur naturally in cells in which two or more gene loci, or alleles at a single locus, are active in the production of different enzyme subunits. Comparisons of the isoenzyme patterns of extracts of different tissues, particularly of the patterns presented by homozygotic and heterozygotic individuals, have provided a fruitful source of insights into the structures of isoenzymes. An opportunity to extend the analysis of hybrid isoenzyme formation in living cells beyond the limits of observation of naturally-occurring examples, into the domain of experimentation, has been provided by the development of techniques of somatic-cell hybridization and culture.

Fusion of somatic cells from different species can be induced to take place in the presence of chemical or viral agents, such as polyethylene glycol or inactivated Sendai virus, to form cell lines with nuclei which contain functional chromosomes from both parent cells. The hybrid cells usually retain the full chromosome complement of one cell, although part of the genome of the other is lost; thus, analogous genes derived from both parent cells are expressed to different extents in the daughter cells and their respective products can be detected in those cases in which differences in properties exist. Differences in electrophoretic mobility have been established for more than 60 enzymes and isoenzymes, derived either from human genes or from the genes of mouse- or Chinese hamster-cells (Shows, 1977), so that the patterns of enzyme zones obtained by electrophoresis of extracts of human cells fused with cells from one of the

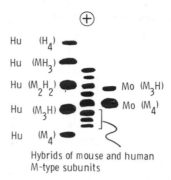

Hybrids of mouse and human
M-type subunits

Fig. 2.6   *Diagram of isoenzymes of lactate dehydrogenase in human (Hu) and mouse (Mo) cells and in hybrid somatic cells. The three possible heteropolymeric isoenzymes formed by association of mouse and human M-type subunits can be distinguished among the isoenzymes expressed in the hybrid cells (Adapted from Harris, 1975).*

other species reflect the genetic composition of the hybrid cells.

The expression of genes determining different isoenzyme subunits in hybrid somatic cells may lead to the appearance in these cells of heteropolymeric isoenzymes (Fig 2.6). The cell-fusion technique thus supplements and extends the experimental production of hybrid iso-enzymes *in vitro*, and the appearance of electrophoretically-separated isoenzyme zones from hybrid cells is open to similar interpretations as to the subunit composition of the respective isoenzymes. The quaternary structures of nearly 50 enzymes and isoenzymes have already been confirmed or determined by the analysis of isoenzyme patterns of hybrid cells (Hopkinson *et al.*, 1976; Shows, 1977).

NON-ISOENZYMIC MULTIPLE FORMS OF ENZYMES

Multiple molecular forms of enzymes which do not originate at the level of the genes determining the structure of the enzymic polypeptides, and which therefore do not fall within the currently recommended definition of isoenzymes, might arise in a number of different ways. Either covalent or non-covalent modifications of the enzyme structure may be involved, resulting in multiple forms which are sometimes referred to as 'secondary isoenzymes'. Although various processes have been shown to account for some of the non-isoenzymic multiple forms of enzymes found in living tissues, and others can be caused to produce multiple forms *in vitro*, the origins and differences in structure of many naturally-occurring multiple forms of enzymes remain unexplained (Fig. 2.7).

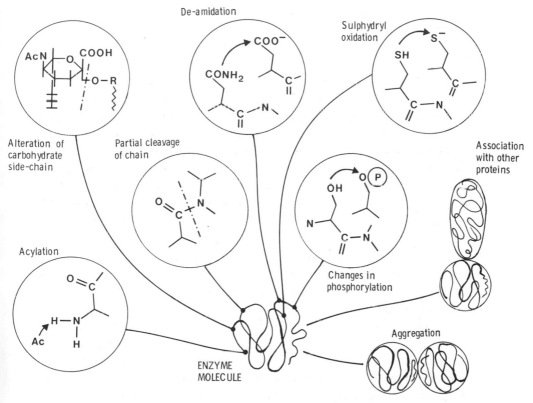

Fig. 2.7 *Diagrammatic summary of post-translational modifications which may give rise to multiple forms of various enzymes.*

Post-translational modifications taking place in living tissue may be as significant in the understanding of certain biological processes, or as useful in clinical diagnosis, as isoenzymic variation arising at the level of the structural gene. Enzyme modification *in vitro* is also not without interest, since it can contribute to the elucidation of enzyme structure, while differential effects of modification may help to distinguish between otherwise similar forms of an enzyme. Furthermore, recognition of possible artefactual modifications is essential for correct interpretation of the observed heterogeneity of enzyme preparations.

Variations in covalent structures

Modification of the structure of the polypeptide chains of certain enzymes may take place within the tissues, with resultant changes in physicochemical properties. Five forms of rabbit-muscle aldolase can be separated by isoelectric focusing. These are due to the intracellular conversion of a single asparagine residue near the carboxyl terminal of the polypeptide chain into an aspartyl residue, causing an increase in net negative charge (Lai *et al.*, 1970). Since the active form of this enzyme is a tetramer, the formation of hybrid molecules of native and modified chains accounts for the observed heterogeneity of the crystalline enzyme. Deamidation has also been identified as the cause of part of the molecular heterogeneity exhibited by certain other enzymes, notably amylase (Karn *et al.*, 1974). Removal of arginine residues from first one and then the other amino terminus of the polypeptide chains of the dimeric alkaline phosphatase molecule of *E. coli* gives rise to two additional forms of this enzyme, with different electrophoretic mobilities from that of the parent molecule (Schlesinger *et al.*, 1975).

The constituent polypetide chains of enzyme molecules may undergo more extensive degradation by proteolytic enzymes, generating multiple enzyme forms. Conversion of a subunit of molecular weight of about 63 000 into a smaller form, molecular weight 58 000, by limited proteolysis is thought to account for the differences between the pyruvate kinases of human erythrocytes and liver. A greater activity of the appropriate protease in liver could account for the preponderance of the smaller molecular form of pyruvate kinase in this tissue (Kahn *et al.*, 1978). Two forms of fructose-1,6-diphosphatase have been identified in rabbit-liver depending on whether the animals are fed or fasted: in the latter state a small peptide appears to have been removed from the enzyme molecule (Horecker, 1975).

Disruption of cells for the extraction and purification of enzymes may present opportunities for interaction between tissue constituents and enzymes which do not exist in the intact cells, causing enzyme modification. An altered form of rabbit-liver fructose-1,6-diphosphatase results from proteolysis under these circumstances, and additional variants of hexokinase from yeast have similarly been shown to be due to the action of proteases on two distinct types of enzyme molecule during extraction from the cells (Gazith *et al.*, 1968). Several erythrocyte enzymes, including adenosine deaminase and acid phosphatase, contain sulphydryl groups which are susceptible to oxidation: in haemolysates, this may be brought about by the action of oxidized glutathione, although in the intact cells this compound is present in its reduced form. Thus, variant enzyme molecules with altered molecular charge may be generated (Hopkinson and Harris, 1969; Hopkinson, 1975). Similar changes are observed in preparations of phosphoglucomutase from human muscle and rabbit tissues and of isoenzyme III of carbonic anhydrase from skeletal muscle (Dawson and Greene, 1975; Carter *et al.*, 1979).

Covalent modification of other ionizable groups present in certain amino acid side-chains of proteins can also produce enzyme forms with altered molecular charge and therefore with distinctive electrophoretic mobilities. An example of this is acetylation of the terminal amino groups of lysine and arginine, which reduces the basicity of these groups and therefore increases the anodal mobility of the protein molecule. Acetylating systems are present in some tissues, but it is not certain that acetylation of protein side-chains takes place *in vivo*. However, acetylation and other covalent modifications of ionizable amino acid side-chains have been used to prepare electrophoretically-altered forms of enzymes such as alkaline phosphates *in vitro* (Moss, 1970a).

Variations in non-polypeptide constituents

Modifications affecting non-protein components of enzyme molecules may also lead to molecular heterogeneity. Many enzymes are glycoproteins, especially those derived from cellular membranes, and variations in the composition of carbohydrate side-chains are a common cause of non-homogeneity of preparations of these enzymes.

The addition of carbohydrate residues to the side-chains of glycoproteins is a function of specific glycosyl-transferring enzymes; therefore, the potentiality exists of differential genetic control, through these enzymes, of the structures of the carbohydrate components of glycoprotein enzymes as well as of non-enzymic glycoproteins. Genetic control of this type is well recognized in the case of the water-soluble, blood group-specific glycoproteins. Glycosyltransferases determined by A or B alleles at the ABO gene locus catalyse the addition of either N-acetyl-D-galactosamine residues or D-galactose residues, respectively, to the terminal positions of the polysaccharide chains of these glycoproteins. Structural variations in the carbohydrate components of multiple forms of enzymes may similarly be manifestations of the differential expression of multiple gene loci or alleles which control glycosyl transferases. However, since structural variants arising in this way would not involve the primary structures of the multiple enzyme forms (i.e. their amino acid sequences), they would not be regarded as true isoenzymes, in spite of their genetic origins.

Evidence has been presented that differences between the electrophoretic patterns of preparations of lysosomal α-mannosidase from certain inbred strains of mice are inherited, and are apparently due to allelism at a single locus on chromosome 5. Expression of different alleles in the livers of various strains of animals results in differences in the amounts of N-acetyl neuraminic acid incorporated into the α-mannosidase molecule, although whether this results from direct genetic control of a specific sialyltransferase or is secondary to other genetically determined processes remains undetermined (Dizik and Elliott, 1977).

Verification that structural differences in carbohydrate side-chains are the cause of multiple forms of glycoprotein enzymes requires a complete analysis of these side-chains, as well as confirmation of the presumed identity of the protein cores of the different forms. Two distinct families of multiple forms of human salivary α-amylase have been identified, one of which consists of glycoproteins whereas the other does not. Superimposed on these differences are a series of further post-translational modifications due to deamidation (Keller *et al.*, 1971; Karn *et al.*, 1974). In this instance the composition of the carbohydrate moiety of the glycosylated family has been determined, but in general definitive structural information is difficult to obtain when enzyme samples are limited in amount and purity. Glycoprotein preparations are typically heterogeneous with respect to charge, probably due to the presence of mixtures of components in which the carbohydrate side-chains are in various stages of completion, and perhaps also to differences between side-chains attached to different sites within the same protein molecule. Furthermore, many glycoprotein enzymes are components of structural elements of the cells in which they occur, such as cell membranes, and the rather vigorous treatments needed to bring these enzymes into solution may result in partial degradation and so contribute to the degree of heterogeneity observed in enzyme preparations, adding to the difficulties of analysis. However, as in the investigation of polypeptide structure, partial information about the nature of differences between the carbohydrate portions of multiple enzyme forms can be derived from specific modifying procedures which result in a selective change in properties.

The carbohydrate side-chains of glycoproteins frequently terminate in N-acetyl neuraminic (sialic) acid residues which are accessible to removal by the enzyme neuraminidase, usually obtained from *Clostridium perfringens* or *Vibrio cholera*. Hydrolysis takes place under mild conditions, e.g., at nearly neutral pH and, although incubation for several hours at 37°C may be needed for complete reaction, there is usually little loss of activity on the part of the substrate isoenzyme. Catalytic properties of glycoprotein isoenzymes are little affected by this treatment. However, the strongly-acidic sialic acid residues have, when present, a marked effect on the net charge of glycoprotein molecules over a wide range of pH, so that their removal results in a considerable reduction in electrophoretic mobility towards the anode, and the effects of digestion with neuraminidase on this property of glycoprotein isoenzymes have been widely studied. Other properties which are influenced by net molecular charge, such as solubility, may also be affected.

Several examples of the use of neuraminidase to differentiate between multiple forms of enzymes have been reported. Human and animal tissue alkaline phosphatases are retarded on electrophoresis after neuraminidase treatment, compared with untreated controls, with the exception of the small-intestinal isoenzyme, which apparently contains no sialic acid residues accessible to the action of neuraminidase (Robinson and Pierce, 1964; Moss et al., 1966). The pronounced heterogeneity with respect to net charge of human kidney alkaline phosphatease also appears to be due to a large extent to the presence of various proportions of sialic acid residues, with only a small proportion of the phosphatase activity of this tissue resistant to the action of neuraminidase (Butterworth and Moss, 1966). Stepwise removal of up to seven sialic acid residues per mole of placental alkaline phosphatase has been achieved by the graded action of neuraminidase (Robson and Harris, 1966). Differing contents of N-acetyl neuraminic acid molecules seem also to account for the many zones of prostatic acid phosphatase separable by starch-gel electrophoresis (Smith and Whitby, 1968). The nature of the intracellular processes which give rise to these differences is generally unknown, but some may represent further examples of the differential genetic control of the degree of sialylation which appears to operate in the case of mouse-liver α-mannosidase.

The presence of neutral carbohydrate molecules as components of multiple forms of enzymes is less easy to determine when pure preparations are not available, since their removal may produce no marked change in properties, while amounts of the carbohydrate molecules released by treatment of impure preparations containing little of the isoenzyme may be undetectable.

The covalent attachment of small molecules or radicals to enzyme molecules may alter their properties so markedly as to generate different molecular forms. Probably the most important examples of this are those multiple forms of enzymes which are due to differences in phosphate content. Two major forms of phenylalanine hydroxylase with characteristic properties prepared from rat liver have been shown to differ in this respect (Donlon and Kaufman, 1980). These forms may represent a further example of the regulation of enzyme activity by phosphorylation and dephosphorylation which was first demonstrated for enzymes of glycogen metabolism (Cohen, 1976).

Non-covalent modifications of enzyme structure

Aggregation of enzyme molecules with each other or with non-enzymic proteins may give rise to multiple molecular forms which can be separated by techniques that depend on differences in molecular size. Formation or dissociation of aggregates of enzyme molecules are common complications of ultracentrifugal analysis, so that it may be difficult to identify the native enzyme forms. High molecular-weight forms of human alkaline phosphatase occur in extracts of tissues such as placenta which appear to be aggregates of the major active form of the enzyme (Ghosh and Fishman, 1968). The mitochondrial isoenzyme of creatine kinase from beef heart can exist in forms differing in molecular size by a factor of three which interconvert in solution depending on the protein concentration and the presence of reducing agents (Hall *et al.*, 1979).

Multiple forms of serum cholinesterase can be separated on electrophoretic media such as starch gel in which molecular size influences segregation of the protein zones. Four catalytically-active cholinesterase components with molecular weights ranging from about 80 000 to 340 000 are found in most sera, with the heaviest component, C_4, contributing most of the enzyme activity. Additional enzyme forms are also present occasionally, but it appears that the principal serum cholinesterase fractions can be attributed to different states of aggregation of a single monomer (La Motta *et al.*, 1970). The 'true' acetyl cholinesterase of related substrate specificity which occurs widely in nervous and other tissues is also capable of existing in several active forms of different molecular size. Globular forms of this enzyme corresponding in molecular weight to a subunit of about 70 000 molecular weight and dimers and tetramers of it can be extracted from vertebrate tissues. These forms also display micro-heterogeneity with respect to their electrophoretic and solubility properties, as do the cholinesterase variants, probably due to their glyco-protein nature.

Asymmetric forms of acetylcholinesterase exist in the electric organs of electric eels and rays, as well as in the motor endplates of rat skeletal muscle, and these forms represent a further level of complexity of quaternary structure (Massoulié, 1980). Their asymmetry results from the presence of protein 'tails' with a collagen-like structure. Each tail is attached to a tetrameric unit of polypeptide subunits by disulphide bridges; disulphide bonds also link pairs of subunits in the tetramer. Two or three of the tailed tetramers can become associated into still larger structures by ionic interactions between the tails in solutions of low ionic strength.

Complex associations between enzymic and non-enzymic proteins or other constituents are characteristic of enzymes which are associated in the tissues with cell membranes or other organelles. Important clues to the mode of assembly and functional significance of such structurally organized enzyme complexes are to be expected from further understanding of the nature of the interactions involved in their formation, and have already been obtained in the case of the post-translational modifications which lead to the differential localization of β-glucuronidase within mouse cells (Lusis and Paigen, 1977), as well as for the cholinesterases. The nature of the interactions which give rise to multiple forms of enzymes in this way can sometimes be inferred, even with impure preparations, from the results of experiments in which the physical properties of the enzymes are modified without loss of the characteristic catalytic activity. The effect of ionic strength on aggregation of the asymmetric forms of acetylcholinesterase is an example of this: the ability to aggregate is abolished by partial cleavage of the tails with collagenase (Massoulié, 1980). Analogous effects are produced by treating γ-glutamyl transferases from various tissues with papain or trypsin, which abolish the tendency of the native enzyme to aggregate. Treatment of preparations of this and other membrane-derived enzymes with detergents or organic solvents frees them from associations with lipids or lipoproteins which may also be a cause of enzyme heterogeneity (Echetebu and Moss, 1982; 1982a).

In the early days of fractionation of mixtures of enzyme molecules by zone electrophoresis, particularly those occurring in serum or plasma, fears were often expressed that the separations, and therefore the apparent number of enzymic components, might be affected by interactions between enzyme molecules and non-enzymic protein molecules. In general, these fears have proved groundless. However, a specific form of interaction between enzymic and non-enzymic proteins *in vivo* has been shown to be the cause of unusual enzyme components seen when some samples of human plasma or serum are fractionated by electrophoresis or chromatography. These components are due to the combination of apparently normal enzyme or isoenzyme molecules with certain plasma immunoglobulins (Chapter 3).

3 Differences in Properties between Multiple Forms of Enzymes

Modern concepts of protein structure, with their emphasis on the uniqueness of the primary structure of each type of polypeptide chain from which follow its specific secondary and tertiary structures, imply that most modifications of the amino acid sequence of the one or more chains which constitute a given protein will result in some observable change in its properties. The extent of such changes is potentially very wide, depending on the nature of the amino acid substitution which distinguishes the modified protein from its parent form: e.g., if an amino acid in or near the active centre of an enzyme is replaced by another, a direct effect on catalytic function is likely, whereas replacement of a residue in a more distant region of the molecule by one with similar characteristics may produce only a minor change in properties.

Even when the nature of differences in primary structure between the isoenzymes of a particular enzyme is known, it is rarely possible to account fully for their differences in properties in these terms, and still less possible to predict the effects of other hypothetical structural modifications. The relationship between structure and function is even less clearly apparent in the case of multiple enzyme forms which do not result from differences in primary structure. Nevertheless, the existence in nature of a wide variety of multiple forms of enzymes provides opportunities to study and extend knowledge of the relationships between the structures and properties of enzyme molecules which cannot at present be matched in the laboratory.

The differences in properties between members of a family of isoenzymes, or between related multiple forms of enzymes in general, can be demonstrated by a wide range of experimental techniques, and provide the basis of methods for the analysis of isoenzyme mixtures.

39

CATALYTIC DIFFERENCES BETWEEN MULTIPLE FORMS OF ENZYMES

Although, by definition, the multiple forms of a particular enzyme all catalyse the same chemical reaction, they are not necessarily identical in their catalytic properties; indeed, the existence of variations between true isoenzymes in these properties appears to be the rule rather than the exception.

Differences in specific activity

Isoenzymes determined by mutant alleles frequently show decreased activity compared with the produce of the usual allele, increased activity being much less common. Differences in enzymic activity *in vivo* associated with the occurrence of variant alleles may result from several factors, besides a real change in catalytic effectiveness of the isoenzyme molecules controlled by the alleles. A reduced rate of synthesis or an increased rate of degradation of a mutant enzyme may account for its lower activity *in vivo*. These alternatives cannot readily be distinguished from a change in specific activity unless an independent means of estimating the amount of protein synthesized by the mutant gene is available. This can be achieved by immunochemical titration, for example, in those cases in which the gene product retains its antigenic identity.

A greater than normal rate of synthesis of one variant of glucose-6-phosphate dehydrogenase, G6PD Hektoen, has been demonstrated (Dern *et al.*, 1969) and the higher activity of serum cholinesterase in the sera of members of two families appeared to be accounted for by an increased synthesis of enzyme protein (Delbrück and Henkel, 1979). In general, however, variant enzymes may be synthesized at the same rates as the normal proteins, so that reduced enzymic activity may be due either to the presence of an isoenzyme of lower than normal specific activity, or to an increased rate of degradation of the gene product. A variant form of the inherited disease galactosaemia is due to synthesis of an isoenzyme of the enzyme galactose-1-phosphate uridylyl transferase (also called hexose-1-phosphate uridylyl transferase) which has only half the usual specific activity, although it is produced in normal quantities (Tedesco, 1972). The purified M_4 and H_4 isoenzymes of lactate dehydrogenase differ in catalytic centre activity, the M_4 form being almost twice as active as the H_4 isoenzyme (Pesce *et al.*, 1967).

Differences in reaction with substrates

Dependence of velocity on substrate concentration

The dependence of the velocity of an enzymic reaction on substrate concentration is expressed for the majority of enzymes by the Michaelis–Menten equation, $v = Vs/(s + K_m)$, in which v is the initial rate of reaction, s the concentration of substrate, V the maximum velocity and K_m the Michaelis constant. Exceptions to the hyperbolic relationship between v and s described by the equation occur in the important category of allosteric enzymes, for which a sigmoid relationship is observed under certain conditions, while for many enzymes deviations from Michaelis–Menten kinetics are seen at high substrate concentrations in the form of inhibition by excess substrate. Also, since the majority of enzymic reactions involve more than one substrate, the dependence of v on the concentration of one substrate is influenced by that of the second substrate in a manner related to the reaction mechanism. Nevertheless, studies of velocity–substrate relationships have occupied an important role in isoenzyme studies, both in the identification and characterization of multiple enzyme forms and in attempting to infer their possible physiological significance.

Comparison of the value of V for different isoenzymes is useful only if an independent estimate is available of the amount of each isoenzyme present, so that their specific activities can be compared. On the other hand, K_m, the substrate concentration at which the observed reaction velocity is half V, is usually independent of enzyme concentration and this parameter is a valuable indication of differences between isoenzymes in affinity for their substrates.

Michaelis constants are usually dependent on pH, ʃ
influenced by other factors such as temperature, the nature o
ionic strength, the presence and concentration of cofactors an
mentioned, the concentration of a second substrate in some tʌ.ᴜ–ꜱᴜʙꜱtrate
reactions. Careful standardization of experimental conditions is therefore
essential for reliable comparisons of K_m values of isoenzymes, especially as
the values may differ only slightly.

Differences in Michaelis constants have been reported for the members of several sets of isoenzymes determined by allelic genes (Sutton and Wagner, 1975). Although less common isoenzymes generally show a reduced affinity for their substrates, i.e. increased K_m values, compared with the more usual form, variants of glucose-6-phosphate dehydrogenase have been described which have increased substrate-affinity, as well as forms which show decreased substrate-binding. The range of variation of K_m values for glucose-6-phosphate in this family of mutant enzymes is of the order of one third to four times the value for the common isoenzyme. Michaelis constants for the coenzyme NADP also differ between some variants.

Isoenzymes determined by multiple gene loci typically differ in their Michaelis constant. The value for the tetramer composed of H-subunits of human lactate dehydrogenase is approximately one tenth of that for the M_4 homopolymer with pyruvate as substrate, and rather less than half in the reverse reaction in which the substrate is lactate. Heteropolymers made up of both H and M subunits have intermediate Michaelis characteristics. The H_4 isoenzyme (LD_1) is inhibited by excess pyruvate to a greater extent than the M_4 isoenzyme (LD_5) when measurements are made under certain conditions (Plagemann et al., 1960).

Other examples of isoenzymic products of multiple gene loci which exhibit differences in Michaelis constants include the MM and BB forms of creatine kinase, the A, B and C isoenzymes of aldolase and the genetically-distinct cytoplasmic and mitochondrial isoenzymes of such enzymes as aspartate aminotransferase and malate dehydrogenase (Eppenberger et al., 1967; Horecker, 1975; Fleisher et al., 1960; Grimm and Doherty, 1961).

As well as differing in K_m values and in response to high substrate concentrations, isoenzymes may differ in inhibition by a product of the reaction: e.g. mitochondrial aspartate aminotransferase is more susceptible to inhibition by oxaloacetate accumulating during the course of the reaction than the cytoplasmic isoenzyme, with the direction of reaction usually chosen for measurement of aminotransferase activity (Boyd, 1961). Since oxaloacetate is also a substrate of the reverse reaction, this corresponds to a greater inhibition of the mitochondrial isoenzyme by an excess of this substrate when the reaction proceeds in the other direction. There is a 40-fold difference in Michaelis constants of 'tissue' and 'serum' variants of human adenosine deaminase (Ellis and Goldberg, 1970). The origins of the multiple forms of this enzyme are unknown.

Apparent identity of Michaelis constants has been used to support the view that some multiple forms of enzymes separable by charge-dependent means do not represent distinct isoenzymes, but are more probably attributable to such phenomena as aggregation of a single type of enzyme or binding of enzyme molecules to non-enzymic proteins. This approach was employed, for example, in the case of the minor zones of alkaline phosphatase seen after starch-gel electrophoresis of human tissue extracts (Moss and King, 1962).

When two or more isoenzymes with different Michaelis constants act on a single substrate, non-linear plots are obtained with the various transformations of the Michaelis–Menten equation (e.g. the double-reciprocal plot of $1/v$ against $1/s$) which typically yield straight lines with single enzymes. Non-rectilinear plots of data according to the transformed Michaelis–Menten equation are therefore a useful indication of heterogeneity of enzyme preparations acting on a single substrate and the possible existence of kinetically-distinct isoenzymes. However, when differences in K_m values are small and a relatively limited range of substrate concentrations is used, deviation of the plot from a straight line may be imperceptible, giving an apparent Michaelis constant intermediate between the extreme values for the isoenzymes composing the mixture.

Numerous attempts have been made to exploit differences in kinetic properties between isoenzymes in order to estimate the respective contributions of individual isoenzymes to the total activity of mixtures of the various forms. For example, because of the differences in Michaelis constants of the LD_1 and LD_5 isoenzymes of lactate dehydrogenase, the ratio of activities at high and low concentrations of pyruvate will depend on the relative proportions of these two isoenzymes in the enzyme sample: with pyruvate concentrations of 1.2 and 0.15 mmol l^{-1} pyruvate, a ratio of approximately 0.4 is observed with extracts of rabbit tissues such as heart or erythrocyte rich in LD_1, compared with 1.5–2.0 for liver or skeletal muscle extracts (Plagemann *et al.*, 1960). Similarly, differences in inhibition by excess substrate of lactate dehydrogenase in extracts of human heart or liver tissue are such that the activity of the heart enzyme is reduced by approximately 60% by increasing the pyruvate concentration from 0.34 to 5 mmol l^{-1}, whereas that of the liver enzyme falls by only 18% (Bernstein, 1977). Although ratios of activities at two substrate concentrations broadly reflect the relative preponderance of the H and M subunits of lactate dehydrogenase in enzyme preparations, clearly-distinct ratios are not observed when unfractionated tissue extracts or serum samples are analysed, since these may differ to a comparatively small extent in the proportions of heteropolymeric isoenzymes which they contain, as well as in their content of the H_4 and M_4 homopolymers. Determination of activity-ratios at two concentrations of pyruvate is therefore an insensitive analytical procedure for multi-component lactate dehydrogenase isoenzyme systems, as are analogous methods for the analysis of mixtures of other heteropolymeric isoenzymes.

Analysis of isoenzyme mixtures by differences in substrate affinities is much simplified when only two components are present. The relative proportions of the cytoplasmic and mitochondrial isoenzymes of aspartate aminotransferase in tissue extracts have been estimated by measuring activity with low aspartate and high 2-oxoglutarate concentrations, conditions which favour the mitochondrial isoenzyme, whereas at high concentrations of both substrates activity is due to the two isoenzymes (Fleisher *et al.*, 1960). In this method, discrimination is aided by determining the activity at low aspartate concentration at pH 6.0, rather than at pH 7.4 as is usual at high concentrations, thus taking advantage of differences in pH-dependence of the mitochondrial and cytoplasmic isoenzymes.

With substrate analogues

The degree of specificity of enzymes for their substrates shows a wide variation, from absolute specificity at one extreme, in which the enzyme is completely inactive towards all compounds other than a unique substrate, to examples in which the only requirements for catalysis are the presence of a particular chemical grouping or type of bond in the putative substrate molecules.

When substrate specificity is less than absolute, isoenzymes are found frequently to differ considerably, both quantitatively and qualitatively, in their reactivity towards substrate analogues. Differences are most marked among members of families of enzymes with group- or bond-specificity, such as the non-specific phosphatases or carboxylic esterases, where they contribute to the difficulties of drawing a distinction between multiple forms of a single enzyme and a group of individual enzymes with overlapping substrate specificities (see Chapter 1). Nevertheless, the characterization of the many multiple forms of non-specific carboxylic esterases in human and other tissues has depended heavily on elec-trophoretic separation of enzyme zones, followed by comparison of the patterns obtained with various synthetic ester substrates, and this constitutes one of the earliest examples of the use of this approach to the study of multiple enzyme forms (Hunter and Markert, 1957). In the case of ali-esterases, for example, alternative substrates consist typically of esters with acyl components of various chain lengths, esterified with α- or β-naphthols, liberation of which by hydrolysis can readily be detected by fluorescence, or by coupling with diazonium salts to produce coloured dyes. Derivatives of indoxyl can be used in a similar manner, air oxidation of indoxyl to indigo eliminating the need for a further chromogenic reaction (Hunter and Burstone, 1960). Changes in the relative intensities of the various enzyme zones with different substrates can be used to assess substrate specificity. More reliable quantitative estimates of the relative reactivity of esterase preparations towards various substrates can be obtained by measuring rates of hydrolysis in solution, after separation of the enzyme forms, e.g. by preparative zone electrophoresis.

The various multiple forms of non-specific acid and alkaline phosphatases are also amenable to characterization by the use of alternative substrates, since almost all orthophosphate esters are hydrolyzed by these enzymes. In the case of human alkaline phosphatases, the only structural requirements for a potential substrate are the presence of a terminal orthophosphoric acid radical, two hydroxyl groups of which are unesterified; thus, inorganic pyrophosphate or polyphosphates such as ADP or ATP are cleaved by alkaline phosphatases, orthophosphate groups being removed sequentially from polyphosphate substrates (Moss and Walli 1969). Consequently, many studies have been made of the relative activities of both acid and alkaline phosphatases towards various derivatives of orthophosphoric acid.

Among acid phosphatases from various tissues, differences in the relative rates of hydrolysis of α- and β-glycerophosphates by the enzymes from erythrocytes and spleen were demonstrated as early as 1934, and further differences with respect to these substrates and phenyl phosphate were later shown for prostatic acid phosphatase as well as for the other two enzymes (Davies, 1934; Abul-Fadl and King, 1949). Investigations of the substrate specificity of human acid phosphatases have continued to the present day, because of the clinical requirement for methods of assay with high specificity for the prostatic enzyme (Chapter 7), and additional phosphate esters studied include *p*-nitrophenyl phosphate, α- and β-naphthyl phosphates, and phenolphthalein and thymolphthalein monophosphates (Babson *et al.*, 1959; Roy *et al.*, 1971).

Differences in relative rates of hydrolysis of alternative substrates by tissue-specific forms of alkaline phosphatase are well established. As with many other criteria which have been applied to this group of enzymes from human and other animal tissues, relative activities with various substrates tend to separate the alkaline phosphatases into two main categories, one comprising the enzymes from such tissues as bone, liver or kidney, the other consisting of the placental and small-intestinal phosphatases (Fig. 3.1). Thus, with β-glycerophosphate as a reference substrate, *p*-nitrophenyl phosphate is a relatively poor substrate for the human intestinal isoenzyme while adenosine monophosphate is rapidly hydrolysed, a reversal of the patterns of activity found for the enzymes of

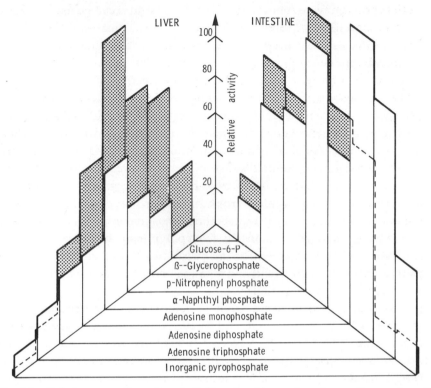

Fig. 3.1 *Relative rates of reaction of alkaline phosphatases from human liver and small intestine with several substrates at pH 9.5, without (open bars) and with (shaded bars) added magnesium ions. The activities of each isoenzyme are related to a value of 100 for hydrolysis of p-nitrophenyl phosphate in the presence of magnesium (From Moss, 1979. By permission of the Chemical Society, London).*

liver, kidney and bone (Landau and Schlamowitz, 1961). Placental alkaline phosphatase is relatively more active towards β-glycerophosphate compared with phenyl phosphate than kidney phosphatase, though with this pair of substrates the intestinal and kidney isoenzymes are rather similar in their relative activities (Ahmed and King, 1960). Dissimilarities in substrate specificity also extend to pyrophosphate substrates such as inorganic pyrophosphate, ADP and ATP, with relatively more rapid hydrolysis of these compounds being effected by intestinal alkaline phosphatase than by non-intestinal isoenzymes when phenyl orthophosphate or a derivative of it is the reference substrate (Moss *et al.*, 1967; Eaton and Moss, 1967). An extensive investigation of the substrate specificities of alkaline phosphatases from rat tissues also demonstrated marked differences between intestinal and other phosphatases in this respect, particularly with regard to the relative sensitivity of o-carboxyphenyl phosphate to hydrolysis by the intestinal enzyme, compared with the resistance of this substrate to attack by the enzyme from liver (Fishman *et al.*, 1962).

The isoenzymes of lactate dehydrogenase differ in their relative activities towards higher homologues of their natural substrate, $L(+)$-lactic acid, and this property has also been exploited as an aid to analysis for clinical purposes. The hydroxy-derivatives of butyric, caproic and valeric acids act as substrates for these isoenzymes, as do the corresponding oxo-compounds in the reverse reaction. Since the reverse reaction is the more rapid, pyruvate and analogues of it are usually chosen in analytical methods based on the differential substrate specificities of lactate dehydrogenase isoenzymes. The ratios of activities with 2-oxobutyrate as substrate to those with pyruvate are of the order of one for the electrophoretically most anodal isoenzyme, LD_1, from human tissues, but less than 0.2 for LD_5, (the most cathodal form), with concentrations of the two substrates of 3.3 and 0.7 mmol l^{-1} respectively and when activities are measured at 25°C (Rosalki and Wilkinson, 1960; Plummer *et al.*, 1963). Thus, sera in which total lactate dehydrogenase is raised due to enzyme release from a tissue rich in LD_1, such as heart muscle, have higher 2-oxobutyrate: pyruvate activity ratios than is the case when LD_5 is released, e.g. from liver as a result of hepatitis. The method has been widely applied in clinical analysis, particularly to give increased sensitivity of detection of myocardial damage (Elliott and Wilkinson, 1961; Konttinen and Halonen, 1962). However, it suffers from the general disadvantages of methods of this type, in that it is insensitive to alterations in the proportions of the intermediate isoenzymes, while the need to determine ratios of two separate measurements magnifies the effects of experimental error. Furthermore, the characteristic ratios of the fast and slow isoenzymes are dependent upon the reaction conditions so that the differences between the isoenzymes may be reduced if these are altered.

Isoenzymes of dehydrogenases are also active to varying degrees when the naturally-occurring second substrates of the reactions which they catalyse, nicotinamide adenine dinucleotide (NAD) or nicotinamide adenine dinucleotide phosphate (NADP), are replaced by various synthetic analogues of these coenzymes. The effects of various substituent groups in either the pyridine or the purine rings have been investigated (Kaplan and Ciotti, 1961). As with experiments in which the first substrate of the reaction is varied, the results of replacing the natural coenzyme with various synthetic analogues are usually expressed as ratios of the activities observed with pairs of alternative coenzymes. Differential effects of coenzyme modifications on the activities of lactate dehydrogenase isoenzymes of several species have been demonstrated (Cahn *et al.*, 1962). With either the 3-thionicotinamide or 3-acetylpyridine analogues as hydrogen acceptors, for example, ratios for the oxidation of lactate by extracts of heart muscle (i.e. a tissue rich in LD_1) from birds, mammals, amphibia and fish all fall within the range of 5–8. Values for skeletal muscle are much more heterogeneous, ranging from as low as 0.2 in fish to more than 2 in man, reflecting the variable isoenzyme composition of this tissue from one species to another. As is the case with many other catalytic properties of lactate dehydrogenase, the heteropolymeric isoenzymes show a graded reactivity towards substrate analogues, between the extremes represented by the respective homopolymers.

Distinct forms of a dehydrogenase acting on compounds which contain vicinal hydroxyl groups, such as glycerol, are produced by *Aerobacter aerogenes* depending on whether the organism is grown on glycerol- or glucose-containing media. These enzyme forms are not physically separable, but can be distinguished by their reactivity with analogues of NAD (Kaplan and Ciotti, 1961). Coenzyme analogues have also been used to differentiate the isoenzymic cytoplasmic and mitochondrial malate dehydrogenases of ox heart and rabbit muscle, and isoenzymes of this enzyme from various tissues of snail, clam and octopus (Grimm and Doherty, 1961; Kaplan and Ciotti, 1961). Malate dehydrogenase also resembles lactate dehydrogenase in that some variation in the structure of the non-coenzyme substrate of the reaction can be tolerated: differentiation between cytoplasmic and mitochondrial isoenzymes, and between homologous isoenzymes from various rat tissues, has been achieved by measuring rates of reduction of mono- and di-fluoro derivatives of oxaloacetate in the reverse reaction catalysed by the enzyme (Kun and Volfin, 1966).

Selective inhibition of isoenzymes

Variations between isoenzymes in their Michaelis constants for particular substrates can be interpreted as reflecting minor differences in the structures of the active centres at which binding of substrate molecules takes place. It is not surprising, therefore, that structural variations of this kind should also cause isoenzymes to differ in their affinities for, and responses to, specific inhibitors, not only when these substances bind to the active centre itself as is the case for competitive inhibitors, but also when other specialized regions of the enzyme molecule are involved in the attachment of the inhibitor and are susceptible to variation. (Specific inhibitory effects of this nature are to be distinguished from irreversible, relatively non-specific inactivation processes which may also reveal differences between multiple forms of enzymes).

Early uses of specific inhibitors to establish organ-specific charac-teristics of enzymes included attempts to discriminate between prostatic and non-prostatic forms of non-specific acid phosphatases, in order to increase the diagnostic value of acid phosphatase measurements in serum in the detection of disease of the prostate (Chapter 7). Some selective inhibitors of acid phosphatases, such as formaldehyde and organic solvents, probably act by denaturation of the enzyme molecules. However, other compounds are reversible inhibitors and, of these, the most useful is dextrorotatory tartrate, a stereospecific fully-competitive inhibitor of the prostatic isoenzyme with an inhibitor constant similar in magnitude to the Michaelis constants of commonly-used substrates. Almost complete inhibition of the prostatic isoenzyme is observed at concentrations of tartrate which have no effect on the red cell enzyme (Abul-Fadl and King, 1949). Although prostate is not the only tissue which contains tartrate-inhibited acid phosphatase, elevation of the activity of the inhibitable isoenzyme in serum is essentially confined to cases of metastatic carcinoma of this tissue.

Several inhibitors have been shown to exert differential effects on the various isoenzymes of lactate dehydrogenase. Isoenzyme 1 is inhibited by sulphite to a greater extent than is the case for isoenzyme 5: with a concentration of the inhibitor of 2×10^{-2} mmol l^{-1}, these two isoenzymes from rat tissues are inhibited to the extent of about 70% and 30%, respectively (Pfleiderer and Jeckel, 1957; Wieland *et al.*, 1959). Lactate dehydrogenases are also inhibited by oxamate and oxalate. Oxamate is a competitive inhibitor of the reduction of pyruvate and a non-competitive inhibitor of the oxidation of lactate, whereas, with oxalate, the modes of inhibition are reversed. Both these inhibitors produce relatively greater effects on the activities of the more acidic isoenzymes compared with their effects on the more basic enzyme forms (Plummer and Wilkinson, 1963).

An early indication of the existence of at least two catalytically non-identical groups of alkaline phosphatases in mammalian tissues was provided by the less pronounced inhibition of intestinal alkaline phosphatase by bile acids than of phosphatases from other sources (Bodansky, 1937). Further confirmation has come from observations with several compounds which inhibit alkaline phosphatases uncompetitively. This type of inhibition, which is uncommon in single-substrate reactions, occurs by combination of the inhibitor with the enzyme-substrate complex, and both V_{max} and the apparent K_m are reduced. Amino acids have long been recognised as potential inhibitors of alkaline phosphatases, some exerting their effects in a non-specific manner which is probably related to interactions with activating or constituent metal ions. However, organ-specific, uncompetitive inhibition was first demonstrated with L-phenylalanine and later with other amino acids (Fishman et al., 1962; Doellgast and Fishman, 1976). Compounds such as L-phenylalanine and L-tryptophan are more inhibitory towards the isoenzymes from placenta, small intestine and some tumours than those from other tissues, whereas the reverse is the case for L-arginine and more particularly for L-homoarginine, a potent inhibitor of alkaline phosphatases from bone or liver. L-Leucine specifically inhibits certain rare variants of placental alkaline phosphatase, as well as forms of the enzyme occasionally detectable in sera of cancer patients (Chapter 6). In all these cases inhibition is stereospecific, the D-isomer of the amino acid being inactive. The broad-spectrum anti-helminthic, levamisole (tetramisole; [(−)-2,3,5,6-tetrahydro-6-phenylimidazo(2,1-b) thiazole hydrochloride]) is also an uncompetitive inhibitor of alkaline phosphatases from human and animal tissues; in this case isoenzymes from tissues other than placenta or small intestine such as bone, liver or kidney, are the more affected (Van Belle, 1976). The derivative of levamisole known as R 8231 [(±)-6(m-bromophenyl)-5,6-dihydroimidazo(2,1-b)thiazole oxalate] is an even more potent inhibitor with similar specificity. None of these inhibitors of alkaline phosphatases is completely specific for the isoenzyme from a particular tissue or group of tissues, nor is inhibition of the more sensitive isoenzymes complete, although activity may be almost entirely abolished by the more effective inhibitors such as levamisole.

Specific inhibitors have played a particularly significant part in the detection and characterization of several forms of serum cholinesterase with reduced catalytic activity which are determined by rare allelic genes (Fig. 3.2). The isoenzyme produced by the so-called 'atypical' allele is less susceptible to inhibitors which contain a positively-charged nitrogen atom than the more common ('usual') form of the enzyme. Several compounds are inhibitors of serum cholinesterase, but the most generally used is dibucaine (Kalow and Genest, 1957). The percentage inhibition at a dibucaine concentration of 10^{-5} mol l^{-1} (the 'dibucaine number') is 80 ± 2 for the typical enzyme and 20 ± 1 for the atypical variant. Differences in

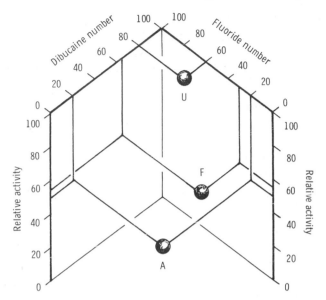

Fig. 3.2   *Differences in relative activity and in degree of inhibition by fluoride ('fluoride number') and dibucaine ('dibucaine number') between 'usual' (U), and 'atypical' (A) and 'fluoride-resistant' (F) allelozymes of human serum cholinesterase.*

the catalytic properties of these two isoenzymes also extend to their Michaelis constants for various substrates, values of these being significantly higher in the case of the atypical isoenzyme. Mutation is therefore thought to have produced an alteration at the site at which the charged group of the substrate or inhibitor is bound, so that affinity for both ligands is reduced.

The usual isoenzyme is inhibited by fluoride and so, to a lesser extent, is the atypical variant. However, the existence of a third, catalytically-distinct isoenzyme was deduced from the observation that fluoride resistance and dibucaine resistance were not invariably correlated. The existence of a 'fluoride resistant' isoenzyme was therefore postulated (Harris and Whittaker, 1961). When inhibition by fluoride at a concentration of 5×10^{-5} mol l^{-1} is expressed as a 'fluoride number', analogous to the dibucaine number, values of the order of 60 are observed for the usual isoenzyme compared with 20 and 30, respectively, for the atypical and fluoride-resistant variants. The dibucaine number of the fluoride-resistant isoenzyme is about 65. Thus, a combination of measurements of inhibition by dibucaine and by fluoride is needed fully to characterize the two variant isoenzymes.

The differences between multiple forms of enzymes in their affinities for specific ligands, such as substrates, substrate analogues or inhibitors, have mainly been investigated by means of experiments in which both enzyme and ligand are in solution. However, these differences can also be explored and exploited for isoenzyme separation by the technique of affinity chromatography, in which the specific ligand is immobilized by covalent attachment to an inert matrix such as cellulose, or gels of dextran or polyacrylamide.

A wide range of ligands can be attached to such matrices. Some of these ligands may react with components of many enzyme molecules and are therefore relatively unspecific: an example is concanavalin A, a plant lectin with a high affinity for α-D-mannosyl, α-D-glucoysyl and sterically similar carbohydrate residues, and consequently for glycoproteins which contain such residues. This and similar lectins have a useful place in separating multiple forms of enzymes containing carbohydrate residues and in comparing this aspect of their structures. However, more structural information and greater selectivity of separation are obtained by the use of ligands which combine with specialized binding sites of particular enzyme or isoenzyme molecules, and several derivatives or analogues of nucleotide coenzymes have been used for this purpose. Cibacron Blue F3G-A coupled to agarose gel ('Blue Sepharose CL-6B'; Pharmacia AB, Uppsala, Sweden) binds a wide range of enzymes with a 'dinucleotide fold' as part of their molecular structures to which nucleotide coenzymes are bound. The most anodal (H_4) isoenzymes of lactate dehydrogenase from many species fail to bind to this type of adsorbent, or are readily eluted with solutions of the nucleotides, nicotinamide adenine dinucleotide (NAD) or adenosine monophosphate. In contrast, M_4 homopolymers are eluted only with reduced NAD. Heteropolymers have intermediate affinities for the ligand (Nadal-Ginard and Markert, 1975). The homodimeric isoenzymes determined by the three human alcohol dehydrogenase loci, ADH_1, ADH_2 and ADH_3, also show differences in affinity for this ligand. In addition, the isoenzymic products of the common allelic genes at the ADH_3 locus differ markedly in their affinities for Blue Sepharose: the $\gamma^1\gamma^1$ isoenzyme is firmly adsorbed, but the $\gamma^2\gamma^2$ dimer is not absorbed, while the $\gamma^1\gamma^2$ hybrid exhibits intermediate affinity characteristics (Adinolfi and Hopkinson, 1978). Adenosine-monophosphate (AMP)-Sepharose can be used to separate the testicular isoenzyme, LD_X, from the lactate dehydrogenase isoenzymes found in other tissues, since LD_X alone is not bound to this absorbent in the presence of 1.6 mmol l^{-1} oxidized NAD (Kolk *et al.*, 1978). Other immobilized coenzymes or their analogues which have been applied in affinity chromatography include pyridoxal phosphate, folic acid and biotin.

Dinucleotides, and other coenzymes, take part in the reaction cycles of many enzymes. Still greater specificity of bonding is obtainable, therefore, with ligands consisting of molecules which are the substrates of enzymes of high substrate specificity, or which are highly-specific inhibitors. A substrate analogue, diamino-caproylphenyltrimethyl-ammonium, bound to agarose gel has been used to separate the usual form of serum cholinesterase from the atypical isoenzyme determined by a rare allelic gene (La Du and Choi, 1975). The atypical isoenzyme can be eluted from the gel with sodium chloride more easily than the usual isoenzyme, as would be expected in view of the comparatively lower affinity for substrates of the atypical form of the enzyme.

Differences between isoenzymes in other catalytic properties

Enzyme activity is frequently enhanced by the presence in the reaction mixture of other substances besides the enzyme and its substrate. Some of these activators are relatively unspecific in their actions; e.g. various divalent metal ions activate a wide variety of different enzymic reactions. The reaction velocity typically shows a hyperbolic, Michaelis–Menten type of dependence on the concentration of activator, and isoenzymes may differ somewhat in their 'activator constants', analogous to the Michaelis constant, derived from this relationship. However, such differences are often small and therefore not useful in isoenzyme characterization.

Certain activators produce their effects by restoring essential groups in the enzyme molecule to their functional state. This is the case for the activation, or reactivation, of creatine kinase by sulphydryl compounds such as reduced glutathione, mercaptoethanol, dithiothreitol, or *N*-acetyl cysteine, which ensure the reduction of thiol groups necessary for catalysis. Some differences have been reported in the extents to which human creatine kinase isoenzymes are activated by various sulphydryl reagents, and a method of estimating the MB isoenzyme in serum has been based on this observation (Rao *et al.*, 1975). However, the differences between isoenzymes are slight, and their molecular basis is unknown.

Isoenzymes frequently differ in their pH optima, and the observation of irregularly-shaped curves for the dependence of reaction velocity on pH can be a valuable indication of the heterogeneity of an enzyme preparation. For example, acid phosphatases from erythrocytes and prostate differ in their pH optima, and so, to a lesser extent, do isoenzymes 1 and 5 of lactate dehydrogenase. However, observed pH optima are markedly affected by such factors as the nature and ionic strength of the buffer solution, the type of substrate when this can be varied, and even, in some cases, the substrate concentration.

Some differential effects of pH may be related to irreversible inactivation of particular enzyme forms, rather than to differences in reversible ionizations, and so may be time-dependent. Careful standardization and control of experimental conditions are necessary, therefore, if differences in pH-dependence are to form a reliable basis for isoenzyme characterization.

PHYSICOCHEMICAL DIFFERENCES BETWEEN MULTIPLE FORMS OF ENZYMES

Differences between multiple forms of enzymes in properties not related to catalytic function (notably in net molecular charge) form the basis of their recognition and separation in the majority of cases. In addition, studies of these properties may allow useful inferences to be drawn about the probable molecular basis of enzyme heterogeneity before the results of definitive structural analysis are available.

Differences in ionic characteristics

About one third of all substitutions of a single amino acid residue for another in a polypeptide chain are likely to involve a change in the ionic charge of the residue at that position, e.g., by substituting a basic amino acid for one which is neutral or acidic, and will therefore result directly in an alteration in the net molecular charge of the polypeptide. However, substitution of one residue for another of different ionic characteristics is not the only way in which differences in the overall charge of a protein molecule may be produced. Since the higher levels of protein structure are determined by the primary structure (i.e. the amino acid sequence), substitution of one amino acid for another may cause a change in molecular conformation such that ionizable residues different from those in the unmodified molecule are exposed to the surrounding medium. The alterations in charge produced by these structural changes may be considerable (Fig. 3.3).

Apart from single amino acid substitutions, other modifications of

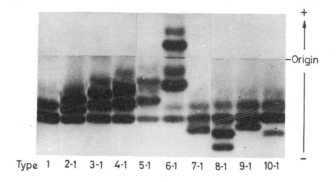

Fig. 3.3 *Zones of phosphohexose isomerase (glucose phosphate isomerase) separated by starch-gel electrophoresis of haemolysates of human erythrocytes. One major homodimeric isoenzyme is present in the homozygous phenotype 1, together with a less intense zone resulting from post-translational modification. The same isoenzyme is also present in all the phenotypes 2–1, 3–1, etc., in which at least two additional zones are present due to the presence of allelic genes at the phosphohexose isomerase locus. The triplet of most prominent zones in each heterozygote consists of the usual homodimer, a homodimer composed of products of the particular mutant allele, and, between these, heterodimers of the usual and mutant gene products. 'Secondary isoenzymes' can be seen in several samples, e.g., accompanying each of the allelozymes in Type 6–1. Each of the mutant isoenzymes probably results from a single amino acid substitution; however, the variation in net charge ranges widely from the most-anodal form of 6–1 to the most-cathodal isoenzyme of 8–1 (From Detter et al., 1968. By permission of Cambridge University Press).*

structure can occur, such as deletion or repetition of a section of the primary structure, with a probable alteration in molecular charge. Thus, although some variations in protein structure arising from the existence of multiple gene loci or alleles may not involve a change in molecular charge, these cases probably make up a smaller proportion of all the possible changes in primary structure than a consideration of the ionization characteristics of amino acids alone would suggest. Similarly, many of the post-genetic modifications which may give rise to multiple forms of enzymes, such as side-chain cleavage, addition of non-protein components, or aggregation, will result in altered molecular charge.

Zone electrophoresis

The potential ability of separation methods based on net molecular charge to detect small differences in composition between protein molecules is therefore very great, and, of these methods, zone electrophoresis has become the most useful single technique in the study of multiple forms of enzymes, which consequently are most often designated by their characteristic electrophoretic mobilities.

As an analytical procedure, zone electrophoresis requires only small samples when combined with suitably sensitive methods of detecting the separated isoenzyme zones. A further advantage of the technique is that the separation achieved is generally unaffected by the degree of purity of the sample being separated. Electrophoresis is therefore particularly valuable for investigating the isoenzyme composition of unpurified extracts of cells or tissues, as well as serum or other body fluids, and in consequence it has been used extensively for genetic and clinical studies. Exceptions to this generalization are occasionally encountered, however, such as the differences in the proportions of minor zones of alkaline phosphatase activity seen with alternative methods of extraction of the enzyme from liver (Moss, 1962) or the changes in mobility of isoenzymes of creatine kinase which result from incubation in serum but not in saline (Cho et al., 1976). Purification may sometimes remove attached ions or small molecules or dissociate isoenzyme complexes, so altering electrophoretic mobility.

Although some of the earliest investigations of the electrophoretic heterogeneity of enzymes were carried out by paper electrophoresis (Baker and Pellegrino, 1954; Wieland and Pfleiderer, 1957; Jermyn and Thomas, 1954) this supporting medium has now largely been superseded by cellulose acetate (Kohn, 1957; Rosalki, 1965; Stagg and Whyley, 1968). Compared with filter paper as a supporting medium, cellulose acetate membranes offer much reduced adsorption of protein zones, so that clearer separation and lower background staining are obtained. Electrophoretic separation is rapid, resolution of isoenzyme zones typically being obtained in one hour or less with potential differences of the order of 25 V cm^{-1}. Thin cellulose acetate membranes are not very absorbent, so that the volume of isoenzyme mixture which can be applied without excessive spreading of the applied zone is limited; similarly, reagents used to demonstrate separated isoenzyme zones may have to be applied by placing filter paper or cellulose acetate moistened with them in contact with the electrophoretic strip, or by incoporating the reagents in an overlay of agar gel, to prevent elution or spreading of the zones.

Agar and agarose gels share some of the advantages of cellulose acetate in that they have low non-specific adsorption of proteins and allow rapid separation of isoenzyme zones. In addition, dried films of the gels are transparent, so that direct photometric or densitometric scanning is possible, while passive diffusion is more restricted in the gels than on cellulose acetate, which contributes to sharper, more clearly-resolved isoenzyme zones. Gels consisting of 0.5–1 % agar or agarose by weight in an appropriate buffer are usually used for electrophoresis, and demons-tration of isoenzyme zones is usually carried out by overlaying the electrophoresis gel with a second layer of gel containing the reagents. The use of purified agarose (the straight-chain polymer fraction of agar) removes much of the variability of results seen with different batches of agar, and also much of the considerable electroendosmosis which is typical of agar-gel electrophoresis because of the acidic sulphate and carboxylate groups present in the agaropectin fraction. Electrophoresis on agar or agarose gels has been used in the study of many isoenzyme systems, including among others lactate dehydrogenase (Wieme, 1959; Blanchaer, 1961; Van der Helm, 1962), alkaline phosphatase (Haije and DeJong, 1963; Hägerstrand and Skude, 1976) and other esterases, aspartate aminotransferase (Boyd, 1962) and creatine kinase (Burger *et al.*, 1964; Deul and Van Breemen, 1964). Agar- or agarose-gel electrophoresis is also particularly useful in conjunction with various immunochemical procedures.

Supporting media such as cellulose acetate membranes or agar gel impose no selective restrictions on the passage through them of molecules of different sizes. In gels formed from partially-hydrolysed starch or by polymerization and cross-linking of acrylamide, however, the pores approximate to macromolecular dimensions, so that larger protein molecules or complexes are retarded during electrophoresis compared with smaller molecules with similar ratios of charge to mass. Restricted diffusion of protein molecules in these gels also helps in maintaining narrow zones during electrophoresis and the subsequent staining process. Starch-gel electrophoresis, first described by Smithies (1955), has oc-cupied a particularly important place in the study of multiple forms of enzymes and continues to do so, as is demonstrated by the extensive compilations of methods which have been published (Shaw and Prasad, 1970; Harris and Hopkinson, 1976), although it has been replaced to some extent by electrophoresis in polyacrylamide gels in recent years.

The numerous applications of starch-gel electrophoresis to studies of multiple forms of enzymes include separations which do not involve differences in molecular size between the forms, as well as separations in which components differing in size are present. In the former case, the sieving properties of the gel do not come into play, but the technique is chosen for its good resolution and compact enzyme zones. Starch-gel electrophoresis can also be combined with electrophoresis on filter paper or other supporting media to give two-dimensional separations. This allows differences between multiple forms of enzymes which are due to differences in net charge to be distinguished from those involving differences in size, e.g. among the multiple forms of cholinesterase in human serum (Harris *et al.*, 1962).

Polyacrylamide gels for zone electrophoresis, formed by polymerization of a mixture of acrylamide with a small proportion of bisacrylamide (*N:N'*-methylene-bis-acrylamide), have virtually no charged groups, so that the electro-endosmosis or adsorption of protein molecules due to ionic effects which may result from the small number of carboxylate groups present in starch gel do not occur. Symmetrical protein zones are therefore usually obtained by electrophoresis in polyacrylamide gels, although aromatic residues in proteins may interact with polyacrylamide to some extent. The Kohlrausch effect has been exploited in polyacrylamide-gel electrophoresis to produce zone sharpening, by the combination of gels of different pore sizes and the use of discontinuous buffer systems (Ornstein, 1964; Davis, 1964; Hjerten *et al.*, 1965).

The reproducibility of properties of polyacrylamide gels and the readiness with which their pore sizes can be varied has led to the introduction of electrophoresis in gradient-pore gels for the estimation of molecular weights (Margolis and Kenrick, 1968). Differences in molecular size can be made the dominating factor in separations in polyacrylamide gel by first treating the mixture of proteins to be analysed with sodium dodecyl sulphate (SDS), then adding a low concentration of SDS (as little as 0.1–0.2 % w/v) to the buffer solutions used for electrophoresis (Shapiro *et al.*, 1967). SDS combines with polypeptides in the ratio of approximately 1.4 g SDS per gram of amino acid residues. Consequently, whatever their original charge properties, different polypeptides thus treated acquire essentially equal ratios of negative charge to mass, because of the attached SDS molecules. When submitted to electrophoresis in polyacrylamide gels, in which resistance to migration varies directly with molecular size, differences in rates of migration of the treated polypeptides reflect their relative sizes.

Reaction with SDS denatures proteins, dissociating polymeric molecules and unfolding their constituent subunits. Because denaturation of enzymes destroys their catalytic activity, enzyme zones separated by SDS electrophoresis cannot be visualized by methods which make use of the reaction catalysed by the enzyme, but only by non-specific methods such as the use of protein stains. Therefore, in contrast to gradient-pore electrophoresis, SDS electrophoresis cannot generally be used to fractionate impure mixtures of isoenzymes which differ in molecular size, although renaturation of enzyme fractions, with recovery of activity after SDS electrophoresis has been achieved in some cases. However, the SDS technique has proved valuable in studies on purified isoenzymes in determining and comparing their subunit compositions.

Proteins can also be dissociated into their component subunits by treatment with concentrated solutions of urea or guanidine hydrochloride, followed by electrophoresis in buffers containing these reagents to separate the subunits. Starch gels can be made in buffers containing up to 8 mol urea l^{-1} and, although setting time is much prolonged, the gels are stronger and more transparent than gels from which urea is absent. The properties of polyacrylamide gel are unaffected by the presence of high concentrations of urea.

Visualization and quantitation of isoenzyme zones. One of the particular advantages of zone electrophoresis in the study of multiple forms of enzymes in that in many cases the separated enzyme zones can be made visible *in situ* in the supporting medium because of their catalytic activity, and the study of isoenzymes is particularly strongly identified with this 'zymogram' technique. The methods used for this purpose are usually derived and adapted from those employed by histochemists to locate enzymes in tissue sections and, as in histochemistry, the presence of enzymes of almost every class can be demonstrated by a suitable choice of substrates or coupled reaction sequences, although ligases and isomerases present difficult methodological problems. The factors which influence the design of methods for locating enzyme zones after electrophoresis are also similar in some respects to those which operate in histochemistry; notably, the need for distinctively coloured reaction products, which are either insoluble or of low diffusibility so that they remain at the sites of enzyme action. The latter requirement, so stringent in histochemistry, can be relaxed to some extent in isoenzyme electrophoresis since a small loss of resolution may be offset by gains of sensitivity or convenience. Similarly, the precautions taken by the histochemist to ensure that reactions competing with, or analogous to, the one under study do not interfere are usually made less necessary by electrophoresis, especially when the enzyme zones are compared under several different conditions of separation.

From the wide range of methods which have been described for the location of specific isoenzyme zones after electrophoresis, certain principles have emerged which are of general applicability. A sequence of oxidation–reduction reactions, by which conversion of the coenzyme NAD^+ to NADH by the action of lactate dehydrogenase was rendered visible by reduction of a tetrazolium salt to an intensely-coloured, insoluble formazan, was introduced by Markert and Møller (1959) to locate the isoenzymes of this enzyme after starch-gel electrophoresis. With the replacement of diaphorase and methylene blue, the intermediate electron carriers of the original system, by phenazine methosulphate (PMS; methylphenazonium methosulphate), this sequence has become not only the means of demonstrating the isoenzyme zones of a wide variety of dehydrogenases, but also of enzymes of other classes with the aid of coupled reactions terminating in the dehydrogenase-catalysed reduction of a tetrazolium salt:

Reduced substrate $\quad$ NAD(P)$^+$ $\quad$ Reduced $\quad$ Oxidized

$\qquad\qquad\qquad\qquad\qquad\qquad\qquad$ PMS $\qquad$ Tetrazolium salt

Oxidized substrate $\quad$ NAD(P)H $\quad$ Oxidized $\quad$ Reduced

$\qquad\qquad\qquad$ Dehydrogenase

For enzymes of relatively low specificity, the chemical nature of the substrate can be varied so that products are formed at the site of enzyme action which are themselves coloured, or which can readily be converted to coloured compounds. Chromogenic substrates include esters of indoxyl or halogenated indoxyls, with orthophosphoric acid or aliphatic carboxylic acids (Sugiura and Hirano, 1977; Hunter and Burstone, 1960), which serve as chromogenic substrates for appropriate esterases. Indoxyl or its derivatives produced by enzymic action are oxidized by air or potassium ferricyanide to indigo dyes. A useful series of esterase substrates can be derived from various naphthols (e.g. 1- or 2-naphthol, naphthol AS, or naphthol AS-MX) or naphthylamines. Although the products of hydrolysis of these substrates are not themselves coloured, the naphthols react readily with stabilized diazonium salts to give intensely-coloured, insoluble dyes (Hunter and Burstone, 1960; Estborn, 1959; Panveliwalla and Moss, 1966).

The ability to vary the nature of the substrate is useful in studies of the substrate specificity of individual isoenzyme zones separated electrophoretically. A series of aliphatic esters of various naphthols has been used in this way in the characterization of mouse-liver esterases separated by starch-gel electrophoresis (Hunter and Burstone, 1960), while the reactivity towards several orthophosphate and pyrophosphate substrates of zones of human liver and small intestinal alkaline phosphatases separated by starch-gel electrophoresis was assessed by conversion of inorganic phosphate to a visible precipitate of lead sulphide (Eaton and Moss, 1967).

Since the normal products of many enzymic reactions are fluorescent (e.g. NADH and NADPH), or become so when synthetic fluorigenic substrates such as esters of aromatic alcohols or amides are used, the appearance of fluorescent zones when the electrophoresis medium is viewed in ultraviolet light is in many cases an alternative method of location of isoenzyme zones which offers greater sensitivity than equivalent chromogenic methods (Moss *et al.*, 1961; Panveliwalla and Moss, 1966; Somer and Konttinen, 1972).

Quantitative estimates of the relative proportions of different iso-enzymes can be obtained by densitometric or fluorimetic scanning of the pattern of isoenzyme zones. Limitations on the accuracy of these methods are imposed by the possibility of non-proportionality between the amount of reaction product in a particular zone and its absorption or emission of light, and, especially for highly-active zones, between the enzymic activity of a zone and the amount of product. In addition, the reaction conditions cannot be matched to the specific catalytic charac-teristics of individual zones, except by the use of replicate separations. Some of these potential errors can be avoided by repeatedly scanning the electrophoretic strip at measured time intervals during the course of the enzymic reaction; for example, if a scanner capable of measuring at wavelengths near 340 nm is used, changes in oxidation of NAD or NADP can be followed without the need for further redox reactions. The successive absorbance readings at the regions of enzyme activity are used to plot progress curves of change of absorbance with time (Wieme, 1959). Substances produced by the action of isoenzyme zones in the supporting medium can subsequently be eluted from the appropriate segments and their concentrations measured by absorptiometry or fluorimetry of the resulting solutions. While errors due to lack of proportionality between the substance and its measured signal can be avoided in this way, the constraint remains that the same reaction conditions must be used for all the enzyme fractions obtained in a single separation. Elution of the isoenzyme zones themselves and subsequent individual measurement of their activities in solution (Moss *et al.*, 1961; Vesell and Bearn, 1961; Pfleiderer and Wachsmuth, 1961) allows the selection of reaction conditions which are optimal for each component of the original mixture, but introduces the possibility of differential recoveries of the components during elution.

The high resolving power of zone electrophoresis, particularly in starch or polyacrylamide gels, has directed attention to the potential value of preparative electrophoresis in isoenzyme studies. However, considerable difficulties have been experienced in devising an entirely satisfactory apparatus for preparative gel electrophoresis, although many designs have been proposed (Jovin *et al.*, 1964; Smith and Moss, 1968; Hodson and Latner, 1971). The differences in net molecular charge between multiple forms of enzymes are more usually exploited on a preparative scale by *ion-exchange chromatography* on substituted celluloses or other carbohydrate polymers (Hess and Walter, 1960; Grossberg *et al.*, 1961; Smith *et al.*, 1968), although the resolving power of this technique is not as great as that of zone electrophoresis on gel media.

Isoelectric focusing

This is the most highly resolving of all separation techniques which make use of differences in the ionization characteristics of protein molecules. It depends on the principle that, when a potential difference is applied to a stabilized pH-gradient, components of a protein mixture will migrate electrophoretically through the gradient until they reach the region in which the pH corresponds to their respective isoelectric points. Since a protein molecule is electrically neutral at its isoelectric pH, no further migration takes place and a series of stationary zones of proteins of different isoelectric points is formed. Continued applications of the potential gradient causes the zones to become more compact, or 'focused'. Although not a new concept, isoelectric focusing has only become practicable in recent years, when the problems of maintaining a uniform, stationary pH gradient have been overcome as the result of the introduction of synthetic carrier ampholytes consisting of polyamino-polycarboxylic acids with isoelectric points covering the range from pH 3.5 to pH 9.5 ('Ampholine'; LKB Produkter AB, Bromma, Sweden). These substances migrate in an electric field to their individual isoelectric points at which they maintain a constant pH by their inherent buffering capacity. The pH gradient is stabilized against disruption by convection currents by a sucrose density gradient, or by polyacrylamide gel or other media which promote little or no electroendosmotic flow.

Among the enzymes and isoenzymes to which these techniques have been applied are L-amino acid oxidase (Hayes and Wellner, 1969), lactate dehydrogenase (Dale and Latner, 1968) and alkaline phosphatase (Smith *et al.,* 1971). However, the very high resolving power of isoelectric focusing is not obtained without some potentially serious disadvantages in isoenzyme separations. The solubility of protein molecules is typically at its lowest at the isoelectric pH, so that precipitation may occur. Stability may also be reduced and loss of activity is not unusual in isoelectric focusing of enzymes.

Differences in stability

The process of denaturation of proteins, by which properties such as solubility are altered strikingly and usually irreversibly by exposure to elevated temperatures or other agents, has for long been recognized as a disruption of the specific three-dimensional structures of the protein molecules. Since denaturation of enzyme molecules is accompanied by a loss of catalytic activity, measurement of enzymic activity under denaturing conditions provides a sensitive indication of the rate and extent of this process. The three-dimensional structures of protein molecules are stabilized by numerous hydrogen bonds and hydrophobic interactions between the amino acids of the polypeptide chains of which they are composed. The conformations of the individual polypeptides, and consequently the number and strength of the stabilizing bonds, are determined ultimately by the primary structures of the polypeptide chains themselves. Even minor changes in critical regions of the structure, extending only as far as replacement of one or a few amino acids, are therefore likely to result in an alteration of the stability of the affected molecule. Comparison of the rates of loss of catalytic activity of multiple forms of enzymes under controlled conditions has become an important method of detecting structural differences between them, and one which can form the basis of quantitative as well as qualitative analysis of isoenzyme mixtures.

Exposure to elevated temperatures or to concentrated solutions of urea or guanidine are most frequently chosen as the denaturing agents. Alternative procedures which may be useful include inactivation by oxidizing or reducing reagents, or by acid or alkaline conditions. Whatever method is chosen, careful control of experimental conditions is essential to ensure that what are often small differences in stability between multiple forms of enzymes are detected and measured reproducibly.

Inactivation by heat

Examples of differences between the multiple forms of particular enzymes in their stabilities to heat are numerous. They include differences between isoenzymes which may be present in the same cellular compartment, such as the isoenzymes of lactate dehydrogenase, as well as between catalytically analogous forms from different intracellular regions, e.g. between the mitochondrial and cytoplasmic isoenzymes of NADP-dependent isocitrate dehydrogenase and between the isoenzymes of aspartate aminotransferase from these locations (Plagemann *et al.*, 1961; Campbell and Moss, 1962; Islam *et al.*, 1972; Wada and Morino, 1964). Isoenzymes determined by different allelic genes frequently have different heat-stabilities. Generally, the more common form of the enzyme is more stable than rarer allelozymes: of 21 allelozymes of glucose-6-phosphate dehydrogenase which have been compared in this respect, 9 were found to be less stable than the common isoenzyme (Harris, 1975). Rare isoenzymes of adenosine deaminase and hypoxanthine-guanine phosphoribosyl transferase with greater than usual stabilities have been reported (Kelley *et al.*, 1967; Hirschhorn *et al.*, 1974).

The extents to which members of a family of isoenzymes differ in their resistance to inactivation by heat range from slight to marked. All the activity of the LD_5 homopolymer of lactate dehydrogenase from rabbit muscle is abolished by heating for 20 minutes at $53°C$, whereas the LD_1 isoenzyme is stable under these conditions. The difference in stability between the MM and BB isoenzymes of creatine kinase from this species is less marked, 75% of the activity of the MM dimer surviving after 20 minutes at $45°C$, compared with 40% of the BB activity. When hybrid isoenzymes made up of unlike subunits can exist, they usually have stabilities intermediate between those of the homopolymers. However, the heterodimer of human enolase, for example, is more stable than the homodimers. Both close similarities and wide differences in rates of inactivation by heat can coexist amongst multiple forms of a single enzyme, e.g. human alkaline phosphatase: placental alkaline phosphatase is completely stable to heating at temperatures up to $70°C$, whereas the enzyme prepared from bone loses half its activity in less than 10 min at $55°C$ at pH 7 (Fig. 3.4). Although by comparison the difference in heat stability between alkaline phosphatases from bone and liver is small, it constitutes one of the few differences in properties by which these tissue-specific forms of the enzyme can be distinguished (Moss and King, 1962).

Temperature coefficients (Q_{10}) for the inactivation of enzyme molecules by heat are high, e.g. for human liver and bone alkaline phosphatases in serum, values of Q_{10} are 40 and 44, respectively (Whitby and Moss, 1975). Therefore, slight variations in temperature between measurements can

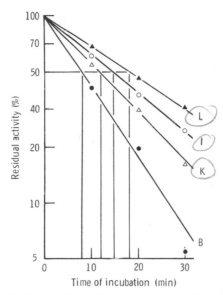

Fig. 3.4 *Inactivation of alkaline phosphatases, partially purified from human tissues by starch-gel electrophoresis, during incubation at 55°C and pH 7. B, bone; K, kidney; I, small intestine; L, liver. The respective half-inactivation times are approximately 8, 12, 15 and 18 min. The scale of the ordinate is logarithmic. Placental alkaline phosphatase is completely stable under these conditions (From Moss, 1979, based on data of Moss and King, 1962. By permission of the Chemical Society, London).*

have effects on rates of inactivation which are sufficiently great to obscure differences between isoenzymes with closely similar stability characteristics. Rates of inactivation of isoenzymes by heat are also generally affected by changes in other factors, notably pH, protein concentration and the concentrations of substrates and cofactors, e.g. the heat-stabilities of isoenzymes of both lactate dehydrogenase and alkaline phosphatase are pH-dependent (Vesell and Yielding, 1968; Moss *et al.*, 1972). Although increases in protein concentration generally have a stabilizing effect on enzyme solutions, the BB dimer of human creatine kinase has been shown to be less stable when incubated at 37°C in solutions containing human or bovine serum albumin (Nealon and Henderson, 1975). The increased stability of enzymes in the presence of their substrates is well known and in some cases a differential effect on isoenzymes is found. For example, the heat stability of the LD_5 isoenzyme is increased in the presence of NADH (Wroblewski and Gregory, 1961).

It is interesting to speculate as to how far the differential thermal stabilities of isoenzymes observed *in vitro* reflect similar characteristics in the living cell and, if stabilities do also differ *in vivo*, whether this contributes to the lower activities of some isoenzymes in certain tissues or in individuals with a particular genetic constitution. The instability of mutant isoenzymes produced by certain allelic genes is thought to contribute to the relative deficiencies in the activities of these isoenzymes in affected subjects. This is most readily apparent when the isoenzymes are present in red blood cells, in which the absence of protein synthesis precludes the possibility that increased enzyme production can compensate for increased destruction.

The greater lability of some isoenzymes compared with others has been suggested as an explanation for their differential rates of disappearance from the circulation, following elevation of enzyme activities as a result of disease. For example, a half-life of approximately two days has been estimated for bone alkaline phosphatase in human plasma (Walton *et al.*, 1975), compared with about seven days for the more stable placental isoenzyme (Clubb *et al.*, 1965), and radioactively-labelled isoenzyme 1 of lactate dehydrogenase disappears more slowly from the blood of rabbits after injection than isoenzyme 5 (Wilkinson and Qureshi, 1976). Therefore, in these cases, the patterns of relative persistence of the isoenzymes *in vivo* are consistent with predictions based on measurements of stability *in vitro*. However, it is not certain that removal of enzymes from the bloodstream is a passive process, in which thermal denaturation plays a major part. Experimental evidence suggests that active removal of a less favoured isoenzyme, as well as increased synthesis of the more favoured form, contributes to the observed distribution of the total enzymic activity of tissues between contributing isoenzymes. In rat liver and skeletal muscle, isoenzyme 5 of lactate dehydrogenase is synthesized at several times the rate occurring in cardiac muscle, but this isoenzyme is also degraded many times less rapidly in liver and skeletal muscle than in heart (Fritz *et al.*, 1969).

Inactivation by urea and related compounds

Concentrated solutions of urea or derivatives of urea such as guanidine and methylurea have a profoundly disruptive effect on higher levels of protein structure. Therefore, differential actions of these compounds on the structures and catalytic activities of multiple forms of enzymes are much used in the analysis and characterization of these forms. The reagents may have several effects on enzyme molecules, although these are not always distinguished in reports of their use.

Exposure to high concentrations of urea ($6-12$ mol l^{-1}) or rather lower concentrations (of the order of 5 mol l^{-1}) of guanidine may cause dissociation of polymeric isoenzymes into their component monomers. This technique has been used to investigate the subunit structures of several families of isoenzymes including, among others, those of lactate dehydrogenase (Appella and Markert, 1961) and creatine kinase (Dawson *et al.*, 1967). In some cases, recombination of monomers to form active enzyme molecules occurs when the urea concentration is reduced, e.g. by dialysis. On the other hand, recovery of catalytic activity may be incomplete or absent, and in these instances derangement of the tertiary structures of the monomers has presumably also occurred. In addition, reversible inhibitory effects of low concentration of urea have been used in distinguishing between multiple forms of enzymes, but these effects seem to be related to differences in catalytic properties rather than in resistance to denaturation.

Inactivation of isoenzymes by concentrated solutions of urea or its derivatives is markedly dependent on temperature and on time of exposure to the reagents. Other variables, such as pH and the presence and concentration of substrates, also influence rates of denaturation; consequently, all these factors must be controlled to ensure reproducible results. Cyanate, itself an inhibitor of some enzymes, forms readily in solutions of urea, which should therefore be freshly prepared.

Other selective inactivation procedures

Many enzymes are metallo-proteins and are therefore inactivated by metal-chelating agents, such as ethylenediamine tetra-acetic acid (EDTA), which remove the constituent metal atoms. As with urea, the irreversible, time-dependent inactivation of metallo-enzymes by EDTA should be distinguished from reversible inhibition, due in the case of EDTA to binding of soluble metal ions which may be activators of the catalytic process. These effects have been studied for alkaline phosphatase isoenzymes, which are zinc-containing proteins, and differences between human bone, intestinal and placental phosphatases exposed to EDTA have been demonstrated (Conyers *et al.*, 1967). Artificial isoenzymes of bacterial alkaline phosphatase have been produced by substitution of other metal atoms for the zinc atoms of the native enzyme form. The native and modified isoenzymes have different catalytic properties (Plocke and Vallee, 1962; Applebury *et al.*, 1970).

Several other agents with effects on the properties of proteins in general have been shown selectively to affect the activities of particular isoenzymes under certain circumstances. Some of these, such as alcohols or other organic solvents which are protein precipitants, feature in early reports of attempts to demonstrate organ-specific characteristics of enzymes. Prostatic acid phosphatase is more susceptible to inactivation by 40% v/v ethanol than the enzyme from some other tissues (Kutscher and Wörner, 1936; Herbert, 1944), while formaldehyde strongly inhibits erythrocyte acid phosphatase but not the prostatic isoenzyme (Abul-Fadl and King, 1948). Ethanol (20% v/v) in acid solution precipitates a greater proportion of liver or placental alkaline phosphatase than of the bone or small-intestinal isoenzymes from solution, e.g. in serum (Peacock *et al.*, 1963). This differential effect seems to be related to the contribution of *N*-acetyl neuraminic acid residues to the net negative charge of the enzyme molecules, since it is largely abolished by previous incubation of the isoenzymes with neuraminidase (Samuelson and Moss, 1978). Exposure to extremes of pH causes denaturation of proteins. Comparison of rates of inactivation of the liver and small intestinal isoenzymes of human alkaline phosphatase at pH 2–4, 0°C, shows the latter isoenzyme to be more resistant to this treatment (Scutt and Moss, 1968).

Because of their differences in primary structure, true isoenzymes (i.e. genetically-distinct enzyme forms) may differ in their resistance to attack by proteolytic enzymes. Different forms of fructose-1,6-diphosphatase with similar catalytic properties are found in the livers of fed and fasted rabbits, that from fasted animals being more susceptible to proteolysis by subtilisin (Horecker, 1975).

Differences in molecular size

When differences in structure between multiple forms of an enzyme are limited to one or a few amino-acid residues, or even if several amino acids are involved, the resulting differences in molecular weight are negligible compared with the molecular weights of the whole molecules. Thus, genetically-determined multiple enzyme forms, in which such alterations are the basis of differences in structure, are typically not separated by techniques which depend on differences in molecular weight or size. On the other hand, aggregation of enzyme molecules, with each other or with non-enzymic proteins or non-protein components, is an important non-genetic cause of enzyme heterogeneity and one which leads to multiple forms differing significantly in molecular size. Post-translational pro-teolysis of polypeptide chains may also be a cause of differently-sized forms of enzymes. Separation methods which exploit these differences are correspondingly important in the investigation of this category of enzyme heterogeneity.

The analytical ultracentrifuge of earlier studies has largely been replaced by methods which permit the identification of separated enzyme fractions by their characteristic catalytic activities. Apart from gel electrophoresis, and especially those modified forms of it in which differences in molecular size are the dominant factors in separation (gradient-pore electrophoresis or SDS-gel electrophoresis), the most useful technique for the resolution of enzyme mixtures according to size is gel filtration. Although the resolution of gel filtration is not equal to that of methods based on gel electrophoresis, it can be applied on a preparative scale.

IMMUNOCHEMICAL DIFFERENCES BETWEEN MULTIPLE FORMS OF ENZYMES

Like other proteins, enzymes stimulate the production of antibodies in species other than those in which they originate, and not infrequently multiple forms of an enzyme are themselves antigenically distinct, giving rise to specific antisera. The antigenicity of proteins is largely due to the topography of relatively small regions of the molecular surface. Non-polar and aromatic amino acid residues appear to be particularly important constituents of antigenic sites, but antigenic specificity is also influenced by the spatial relationships determined by higher levels of protein structure.

The polypeptide subunits from which some isoenzymes are formed may carry different determinants, and an antiserum raised against an iso-enzyme homopolymeric with respect to one type of subunit typically does not cross-react with a different homopolymer. When heteropolymeric isoenzymes can occur, such an antiserum cross-reacts with them to extents which depend on the proportion of the antigenic subunits that they contain. The antigenic characteristics of the dimeric isoenzymes of human creatine kinase exemplify relationships of this kind. Antisera raised in rabbits to either purified MM or BB creatine kinase isoenzymes were each found to be specific for their respective homologous antigens, with no cross-reaction with the heterologous isoenzyme. Both antisera interacted incompletely with the hybrid MB-creatine kinase (Jockers-Wretou and Pfleiderer, 1975). Similarly, an antiserum prepared by immunizing goats with the MM isoenzyme of the human enzyme almost completely inhibited the activity of this isoenzyme but had no effect on the BB dimer. The mixed MB dimer was inhibited to the extent of rather more than 50% (Neumeier *et al.*, 1976; Gerhardt *et al.*, 1977).

The antigenic properties of the tetrameric lactate dehydrogenase isoenzymes are in many respects similar to those of creatine kinase, in that antisera raised against either of the homopolymeric isoenzymes, LD_1 (H_4) and LD_5 (M_4) do not cross-react with the heterologous antigen, and display various degrees of cross-reactivity with the heteropolymers. However, the dissociated monomers of either type from bovine isoenzymes do not combine with rabbit antibodies, and the antigenically-active species is therefore thought at least to be dimeric, and possibly tetrameric (Markert and Appella, 1963), thus providing evidence of the need, in some cases, for a quarternary structure as the basis of antigenicity. Furthermore, evidence from absorption of rabbit antisera raised against a mixture of isoenzymes with one of the homopolymeric forms suggests that some antibodies recognize not just one of the dimeric pairs, HH or MM, but the three combinations, HH, MM and HM (Markert and Appella, 1963).

The immune systems of animals of different species may respond to distinct antigenic determinants when challenged with a particular protein molecule, so that isoenzymes which are antigenically indistinguishable in one species may give rise to different antibodies, and thus to isoenzyme-specific antisera, in animals of a species in which other antigenic determinants are recognized. Similarly, members of a family of iso-enzymes from one species may not be antigenically distinct, whereas the corresponding isoenzymes from a different species may stimulate the production of specific antibodies. Examples of interspecies differences in antigenicity are to be found amongst α-amylases from various tissues. Antisera to human pancreatic amylase raised in rabbits did not dif-ferentiate between this isoenzyme and human salivary amylase (Carney, 1976), whereas a rabbit antiserum to porcine pancreatic amylase ap-parently did not react with amylases from other tissues of the pig (Ryan *et al.*, 1975).

In many cases, interaction between the enzyme antigen and its specific antibody does not involve the active centre of the enzyme, which remains accessible to the substrate. Therefore, the enzyme–antibody complex is catalytically active. In some instances, the reaction between enzyme and antibody abolishes or greatly reduces catalytic activity, most probably because of steric hindrance.

Besides those already mentioned, examples of the use of immunochem-ical methods in the characterization of multiple forms of enzymes are numerous. The value of these methods is particularly apparent in the investigation of inherited enzyme variants. In these cases, quantitative changes in the level of a particular enzyme activity may occur without significant alterations in various parameters such as Michaelis constant or pH-dependence, which are characteristic of the catalysed reaction. Indeed, the level of activity may be so low that these properties cannot be measured, while other characterization procedures are equally impracticable. When the product of the mutant gene retains the antigenic determinants of the usual isoenzyme it can be identified immunochemically, offering the possibility of purification and further characterization, as well as measurement of its amount.

The use of immunochemical assays to demonstrate the production by rare allelic genes of isoenzymes with reduced or increased specific activity compared with the products of the corresponding usual genes has already been mentioned. Some alleles responsible for certain enzyme-deficiency states and inborn errors of metabolism give rise to products (termed cross-reacting materials) which, while totally devoid of enzymic activity, nevertheless retain antigenic determinants which are recognized by antisera raised against the normal isoenzyme.

The presence of antigenic determinants in polypeptides produced by the mutant allele responsible for Tay-Sachs disease has been shown by reaction with an antiserum raised against hexosaminidase A or the α-subunits of the enzyme, each of which had been cross-linked with glutaraldehyde (Srivastava *et al.*, 1979). This disease is caused by a deficiency of the A isoenzyme of hexosaminidase because of a failure of production of the α-subunits of which it is partially composed (Fig. 5.3). Cross-reacting materials cannot be demonstrated in the tissues of affected individuals with antisera to the native α-subunits or isoenzyme A. It is therefore inferred that the cross-linking process exposes antigenic determinants in the α-subunits which are concealed in the native protein, and that the alterations in structure brought about by mutations similarly uncover these determinants (Srivastava *et al.*, 1979).

These and other ineffective subunits or inactive 'isoenzymes' can be regarded as the most extreme examples of functional alterations resulting from genetic mutation.

Immunochemical methods are also useful in elucidating the nature of changes in the activities of specific isoenzymes in tissues which may occur in the course of disease, or in response to specific inducing agents. Two apparently different processes resulting in an increased alkaline phosphatase activity have been distinguished with the aid of immunochemical titrations. Ligation of the common bile duct of the rat is followed by a rapid rise in phosphatase activity in the liver, due to an increase in the activity of the form of the enzyme which is predominant in this tissue. The change in activity was shown immunochemically to be accompanied by an increase in the amount of enzyme protein (Schlaeger, 1975). In contrast, however, an increased amount of enzyme protein could not be demonstrated during the enhancement by cortisone of the activity of a different form of alkaline phosphatase in HeLa cells (Cox *et al.*, 1971). The presence of non-protein components in isoenzyme molecules may introduce additional antigenic determinants, either because these components are themselves antigens, or because they act as haptens when attached to proteins. Of these non-protein components, saccharides are most likely to act as antigenic determinants, as, for example, do the blood-group determining polysaccharides forming part of glycoproteins present at the surfaces of red blood cells. However, some polysaccharide components of glycoprotein enzymes of cellular membranes appear not to be antigenic, perhaps because of their wide distribution in nature.

Interaction of isoenzymes with antibodies may also take place *in vivo* in some circumstances. Several enzymes and isoenzymes have been shown to occur in plasma in the form of high molecular-weight complexes. The non-enzymic components of these complexes are immunoglobulins which can themselves be identified by specific antisera raised against them. The combination between enzyme and immunoglobulin has the characteristics of antigen–antibody reactions; e.g., the complexes are reversibly dissociable. Formation of enzyme–immunoglobulin complexes appears in some cases to be relatively specific for a particular isoenzyme; not surprisingly, perhaps, in view of the frequency of antigenic differences between isoenzymes.

Association of amylase in serum with immunoglobulins gives rise to the condition of macroamylasaemia. The protein combined with the enzyme is either immunoglobulin A or G, and the identity of the immunoglobulin and its presence in the complex can be demonstrated with the aid of appropriate antisera. Some degree of selectivity towards the isoenzymes of amylase has been shown in studies of the nature of the amylase found in naturally-occurring macroamylase complexes, and of the tendency of the dissociated binding globulins to recombine with amylases *in vitro*. In both cases preferential complexing of salivary amylase rather than the pancreatic isoenzyme has been observed (Fridhandler *et al.*, 1974, 1974a). Association of alkaline phosphatase with immunoglobulin G in some sera seems to be specific for the immunologically-similar forms of the enzyme from liver and bone, and the antigenically-distinct isoenzyme from small intestine is not bound (Nagamine and Okhuma, 1975; Crofton and Smith, 1978). Complexes of creatine kinase with plasma immunoglobulins have also been reported (Jockers-Wretou and Plessing, 1979).

Immunoglobulins A and G have also been identified in complexes with the isoenzymes of lactate dehydrogenase. The complexes may show considerable heterogeneity: ranges of molecular weights from 300 000 to 1 000 000 and of isoelectric points from pH 6.3 to 7.2 were found in complexes from the serum of a single patient (Biwenga, 1973; 1977). Formation of complexes involving immunoglobulin A appears to be relatively specific for isoenzyme 2 and to a lesser extent isoenzyme 3 of lactate dehydrogenase (Nagamine, 1972; Markel and Janich, 1974). This apparent specificity is somewhat surprising if the interaction between the isoenzyme and immunoglobulin is due to antigen recognition by the immunoglobulin, since it would be expected that the effective determinants would be present also in at least some other lactate dehydrogenase isoenzyme molecules, in view of their shared polypeptide subunits. An explanation may be that these antibodies to 'self' antigens may be directed against determinants which are only brought together in the appropriate configuration in the quaternary structure of the LD_2 isoenzyme, whereas antibodies raised in other species recognize characteristic features of H- or M-subunits.

Immunoanalytical procedures useful in studies of multiple forms of enzymes

As well as providing indications of differences in structure which may not be demonstrable by other means, the antigenic specificity shown by many multiple forms of enzymes gives access to a wide range of qualitative and quantitative methods of analysis which offer a high degree of specificity and, in some cases, great sensitivity.

The applicability of immunological methods to isoenzyme analysis depends on the production of specific antisera directed against particular isoenzymes, which in turn requires the availability of suitable purified isoenzyme preparations to serve as antigens. Antibodies against a particular antigen frequently vary considerably in their specificity and affinity for the antigen from one preparation of antiserum to another. Variations in specificity from one batch of antiserum to another, or heterogeneity of antibodies in this respect, either within a pool of antiserum from several sources or even with in a serum from a single animal, can have a marked effect on the extent to which the antiserum cross-reacts with molecules with a close structural relationship to the immunizing antigen, e.g. other members of a family of isoenzymes. The specificity of the antiserum for the chosen antigen can be improved by selectively enriching the required antibodies by a specific separation process such as affinity chromatography, or by adsorbing cross-reacting antibodies with the potential antigens for which they have an affinity; an example of the latter approach is the treatment of antiserum to human intestinal alkaline phosphatase with the antigenically-similar placental phosphatase, to increase its specificity for the intestinal isoenzyme (Lehmann, 1975; 1975a). Antisera raised in different animals or at different times generally vary both in the avidity of the antibody for the antigen and in titre, i.e. the concentration of antibodies in the antiserum.

The formation of catalytically-active isoenzyme–antibody complexes is an advantage in various methods of immunochemical analysis which depend on the location of the complexes precipitated in gels. Inhibiting antibodies are useful in some forms of analysis in solution, while radioimmunoassay is entirely independent of catalytic activity.

Immunochemical reactions in solution

In this type of analysis, a solution of the isoenzyme antigen is allowed to react with a specific antiserum and the amount of bound isoenzyme is then estimated. When combination with antibody inhibits catalytic activity, the diminution of enzyme activity is a convenient index of the extent of antibody–antigen reaction. The use of inhibiting antisera is valuable in routine analysis of the isoenzyme composition of large numbers of samples, since the process of combination may be essentially complete within a period of minutes and no separation of bound and free isoenzyme is necessary. Antisera which inhibit the activity of specific isoenzymes of human creatine kinase in serum are widely used for diagnostic purposes (Neumeier *et al.*, 1976; Gerhardt *et al.*, 1977).

Enzyme inhibition is not invariably a consequence of combination with antibody, and indeed complete inhibition probably occurs in only a minority of cases. Frequently, however, conditions can be found by varying the relative proportions of antigen and antibody under which the antigen–antibody complex is insoluble and so can be removed, by allowing it to sediment or by centrifugation. The reaction of enzyme and antibody can thus be demonstrated by the partition of catalytic activity between the sediment and the supernatant. Precipitation of soluble antigen–antibody complexes can be induced to occur in some cases by the addition of a second antibody, specific for the immunoglobulin with which the primary antigen (e.g. the isoenzyme) has combined. This technique has been applied, for example, to the precipitation of soluble complexes of rat-liver alkaline phosphatase with rabbit antibodies, by means of anti-rabbit γ-globulin antiserum raised in goats (Schlaeger, 1975). Precipitation with a second antiserum is also useful in searching for complexes which may form between the anti-isoenzyme antiserum and cross-reacting antigens under conditions which otherwise do not favour precipitation.

Apart from those already mentioned, examples of the use of antigen–antibody reactions in solution as an aid to the differentiation of isoenzymes are numerous, and include their application to lactate dehydrogenase (Nisselbaum and Bodansky, 1959; Plagemann *et al.*, 1960), creatine kinase (Eppenberger *et al.*, 1967), aspartate aminotransferase (Massarat and Lang, 1965), alkaline phosphatase (Schlamowitz and Bodansky, 1959; Boyer, 1963) and other isoenzyme systems (Tedesco, 1972; Goedde and Altland, 1968; Rubin *et al.*, 1971). In some methods sensitivity has been increased by polymerization of the antibody molecules (e.g. by reaction with glutaraldehyde) to render them less soluble, or by coupling the antibody to an insoluble matrix (Lehmann, 1975a; Lee *et al.*, 1978).

Reactions between isoenzymes and specific antisera have been used as a preliminary to zone electrophoresis, to modify the migration of the isoenzymes or to aid in the identification of particular zones. Examples are the electrophoresis of supernatant solutions after the precipitation of isoenzymes of alkaline phosphatase with antisera (Boyer, 1963), and the use of anti-BB-creatine kinase antiserum to distinguish between the BB isoenzyme and a fluorescent artefact of similar electrophoretic mobility (Van Lente and Galen, 1978).

Precipitation reactions in gels

Gels provide a favourable environment for the precipitation of antigen–antibody complexes, by reducing convection currents and thermal agitation which may disrupt the growth of the lattice of antigen and antibody molecules. Diffusion of the antigen and antibody towards each other in gels generates concentration gradients of each which, at some point in their intersection, usually provide the relative concentrations appropriate for precipitation. When this process has begun, migration towards the precipitation zone of further molecules of antigen and antibody increases the amount of precipitated complex. However, the passive diffusion of large molecules through gels is slow, so that the initial reaction between antigen and antibody is retarded and a long period is usually needed to accumulate significant amounts of precipitate. Diffu-

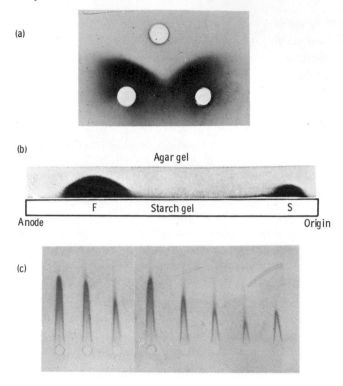

Fig. 3.5   *Examples of three immunodiffusion methods in isoenzyme characterization. In each case an antiserum against human placental alkaline phosphatase was used and precipitates were stained to show enzyme activity.*

(a)   *Double immunodiffusion, showing antigenic identity between normal placental phosphatase (left) and a carcinoplacental isoenzyme (right). The upper well contained antiserum (From Moss, 1979. By permission of the Chemical Society, London).*

(b)   *Immunoelectrophoresis of an extract of human placenta. Enzyme zones (F and S) of different molecular size separated by starch-gel electrophoresis were allowed to diffuse through agar gel towards the antiserum, with which both components react (From Moss, 1973. By permission of Elsevier).*

(c)   *'Rocket' electroimmunoassay of placental alkaline phosphatase in human serum. The activities of the samples ranged from 0.8–12 times the upper limit of normal for total alkaline phosphatase activity in adult serum (From Forman et al., 1976. By permission of Elsevier).*

sion of antigen and antibody through the gel is accelerated in several methods by applying an electric potential.

Positive identification as well as sensitive detection of the precipitates can be achieved in those cases in which the isoenzyme–antibody complex retains its catalytic activity by carrying out in the gel specific staining reactions, such as those used to locate isoenzyme zones after electrophoresis (Fig. 3.5).

Immunodiffusion. In the various applications of this technique, the antibody and antigen move through the gel by passive diffusion. Agarose is now almost invariably chosen to prepare the gel, because of the clarity, purity and mechanical strength of its gels at low concentrations and the absence of negatively-charged groups which may retard the migration of basic proteins, although the pronounced electroosmosis in agar gels can be turned to advantage in some electroimmunodiffusion techniques. The structure of the gels does not present a barrier to the diffusion of large molecules at the concentrations of agarose or other gel-forming materials (agar, starch or acrylamide) usually chosen. In single diffusion methods, one of the reactants, usually the antigen, migrates into a gel which typically contains the antibody. Double-diffusion techniques involve the migration of both components of the reaction towards each other within the gel.

Radial diffusion of isoenzyme antigens into gels containing antisera has been applied to the quantitative assay of isoenzymes (Pfleiderer *et al.*, 1974; Geiger *et al.*, 1975). However, double immunodiffusion according to the Ouchterlony technique (Fig. 3.5) is more useful in the characterization of multiple forms of enzymes, since the appearance of the preciptin lines derived from the simultaneous diffusion of several antigens, or potential antigens, towards a single antiserum can indicate the degree of immunological similarity between the antigens (Lehmann, 1975a; Jockers-Wretou and Pfleiderer, 1975; Uriel, 1963; Foti *et al.*, 1975; Markert and Appella, 1963; Grimm and Doherty, 1961; Tedesco, 1972; Goedde and Altland, 1968; Rubin *et al.*, 1971).

Immunoelectrophoresis. In this technique, which is an important variant of double immunodiffusion, an initial, longitudinal separation of a mixture of antigens by electrophoresis is followed by lateral diffusion of the separated zones, to meet and react with antibody diffusing from a trough cut parallel to the direction of electrophoretic migration (Fig. 3.5). Thus, the antigenic similarities or differences of the components of a mixture can be investigated simultaneously in a system which allows their separation and identification by their characteristic electrophoretic properties. These advantages have ensured the extensive use of immunoelectrophoresis in the study of enzyme heterogeneity (Uriel, 1963; Foti *et al.*, 1975; Grimm and Doherty, 1961; Kaminski, 1966; Goedde and Altland, 1968; Shibuta *et al.*, 1967).

A single gel, typically agar or agarose, is frequently used for both the initial electrophoretic separation and the subsequent immunodiffusion. However, combinations of different gels for the stages of electrophoresis

and immunodiffusion may be useful, particularly when more concentrated gels of starch or polyacrylamide are needed to separate enzyme
zones of different molecular sizes by molecular sieving during electrophoresis, since passive diffusion of large molecules in these small-pore
gels is so slow that they are unsuitable as media for immunodiffusion
processes (Moss, 1973).

Electrically-assisted immunodiffusion. Methods in which diffusion is
accelerated by applying an electric field not only have advantages of
rapidity but also are more suitable for quantitative analysis than
analogous passive diffusion procedures. Both single and double electroimmunodiffusion techniques have been applied to the study of multiple
forms of enzymes. In the former, as in single immunodiffusion without the
aid of an applied electric field, the antigen under study migrates through a
gel containing the stationary antibody, or, much more rarely, antibody
molecules travel through a field of antigen. In double electroimmunodiffusion (often called counter immunoelectrophoresis), both
antigen and antibody molecules migrate under the influence of the electric
field.

Antibody molecules generally have little net charge at alkaline pH
values such as pH 8.6. In contrast, many protein antigens, including
isoenzymes, exhibit appreciable electrophoretic mobility under these
conditions. Thus, the necessary movement of antigen and immobility of
antibody can generally be achieved. When the antigen and antibody have
similar ionization characteristics, it becomes difficult or impossible to
select pH conditions under which their electrophoretic mobilities are
sufficiently different for electroimmunodiffusion to be practicable. In
such cases, useful differences in net charge at a given pH can sometimes be
created by chemical modification of one or other components of the
system (usually the antigen). Needless to say, such modification must not
materially alter the antigenic identity of the molecule, nor should it
destroy the catalytic activity of an enzyme antigen, if this property is to be
used to locate the antigen–antibody precipitate. Acetylation or carbamoylation of the free amino groups of proteins prevents their ionization to
positively-charged NH_3^+ residues, thus increasing the net negative charge
of protein molecules and their electrophoretic mobility towards the
anode. Carbamoylation with cyanate has been used to improve the
migration of isoenzyme C (isoenzyme II) of human carbonic anhydrase
during electroimmunoassay (Nørgaard-Pedersen, 1973).

Rocket immunoelectrophoresis. This is also known as Laurell electroimmunoassay. Antigen molecules are drawn from their wells through the antibody-containing gel under the influence of an electric field. As in passive diffusion, migration continues until antigen–antibody complexes are precipitated (Fig. 3.5). The distance into the gel at which precipitation occurs depends on the antigen concentration, since the concentration of antibody is initially constant throughout the gel; thus, the height of the rocket-shaped preciptin peak is a measure of antigen concentration.

Rocket electroimmunoassay has been applied to the analysis of several enzymes and isoenzymes, including carbonic anhydrase (Nørgaard-Pedersen, 1973), placental alkaline phosphatase (Forman *et al.*, 1976), prostatic acid phosphatase (Milisaukas and Rose, 1972), glucose-6-phosphate dehydrogenase in human leukaemias (Kahn *et al.*, 1975) and several hydrolytic enzymes in a human diploid cell line (Milisaukas and Rose, 1973).

Crossed immunoelectrophoresis. The antiserum used in rocket immunoassay ideally should not cross-react with isoenzymes related to that under study, since identification of a particular isoenzyme by its electrophoretic mobility is not possible in this one-dimensional technique. Crossed immunoelectrophoresis combines the separation and identification of isoenzyme zones given by electrophoresis with the rapid and quantitative immunochemical analysis characteristic of rocket electroimmunoassay (Laurell, 1965). A preliminary electrophoretic separation of the mixture of antigens (i.e. the isoenzymes) is followed by a second migration of the zones in a direction at right angles to their original movement into an antibody-containing gel, usually of agarose, again under the influence of an electric field.

Crossed immunoelectrophoresis has been used to determine the amounts of two α-amylase isoenzymes during germination of barley seeds (Bøg-Hansen and Daussant, 1974). In a study of the reactions of a variety of enzymes in microsomes and plasma membranes from rat liver with antisera to each of these cell fractions, the greater resolving power of crossed immunoelectrophoresis was demonstrated by the recognition of ten antigens with uridine diphosphatase activity in the microsomal fraction by this technique, compared with only three which could be resolved by conventional immunoelectrophoresis (Blomberg and Raftell, 1974; Raftell and Blomberg, 1974). Specific enzymic staining reactions were used to identify the antigen–antibody precipitates in all these investigations.

Counter immunoelectrophoresis. This is an electrically assisted variant of double immunodiffusion in which both antigen and antibody migrate in the electric field. To ensure that they meet, conditions must be chosen under which the antigen and antibody move in opposite directions. Ideally, therefore, the pH of the medium should be between the respective isoelectric pH values of these two components, so that one moves towards the cathode and the other towards the anode. However, the isoelectric points of antibodies, although somewhat higher, are not markedly different from the corresponding values for many isoenzymic antigens, so that this condition is rarely achieved in practice. Nevertheless, the principle of counter immunoelectrophoresis can be applied under circumstances in which the antigen migrates electrophoretically towards the anode while the antibody moves in the reverse direction largely by electroendosmosis.

Two counter immunoelectrophoretic assays have been described for the prostatic isoenzyme of acid phosphatase in human serum (Foti et al., 1978; Chu et al., 1978). Both employ agarose gel as the diffusion medium, with naphthol AS-MX phosphate or α-naphthyl phosphate as substrates for the determination of the enzymically-active immunoprecipitate, the liberated naphthols being coupled with stabilized diazonium salts (Fast Red Violet LB or Fast Red Garnet GBC) to form coloured precipitates.

Radioimmunoassay

A growing number of applications of this specific and sensitive principle to isoenzyme analysis have been described, including the determination of carbonic anhydrase isoenzymes (Headings and Tashian, 1970), type II carboxypeptidase B from human pancreas (Geokas et al., 1974), porcine pancreatic amylase (Ryan et al., 1975), human hexosaminidases A and B (Geiger et al., 1975), the prostatic isoenzyme of human acid phosphatase (Foti et al., 1975; Vihko et al., 1978a), human placental and carcinoplacental (Regan) alkaline phosphatases (Jacoby and Bagshawe, 1972; Iino et al., 1972; Chang et al., 1975), and the M and B subunits, and thus the three isoenzymic dimers, of creatine kinase from human tissues (Roberts et al., 1976; Zweig et al., 1978). All these methods make use of ^{125}I for radioactive labelling of a sample of the isoenzyme, and in most cases the double-antibody technique, in which the antibody-bound isoenzyme is precipitated with an anti-γ-globulin antiserum and recovered by centrifugation, is employed to separate the bound and unbound phases. Anti-isoenzyme antibodies polymerized with ethyl chloroformate and homogenized to fine particles, form an insoluble binding phase in one assay of carcinoplacental alkaline phosphatase, centrifugal sedimentation of the bound fraction of the antigen being further aided by the addition of an homogenized starch-gel suspension (Chang *et al.*, 1975). The bound and free antigen fractions have been separated by a solid-phase technique in some assays, by coating polypropylene tubes with the anti-isoenzyme antiserum (Foti *et al.*, 1975: Jacoby and Bagshawe, 1972). Placental alkaline phosphatase has been assayed by a combined double-antibody and solid-phase technique, in which the anti-γ-globulin antibodies are added in the form of insoluble cellulose particles coated with these antibodies to aid subsequent centrifugal separation of the antibody-bound phase (Holmgren *et al.*, 1978).

Immunocytochemistry

The specific location of isoenzymes in cells and tissues can be approached in some cases by adding specific inhibitors to the reaction systems used for the histochemical location of particular enzymes; e.g. L (+)-tartrate has been used in this way to improve the isoenzymic specificity of staining for acid phosphatase. However, new possibilities of sensitivity and specificity of isoenzyme location have been introduced by the availability of isoenzyme-specific antisera in the techniques of immunocytochemistry (Van Noorden and Polak, 1981).

Attachment of the specific antibody to the isoenzyme antigen can be visualized directly if the antibody is labelled, e.g. with a fluorescent marker. Indirect methods have also been developed, in which the bound antibody is further combined with a second anti-immunoglobulin antibody. The second antibody is itself conjugated with peroxidase and its binding is detected by the peroxidase-catalysed reaction. Indirect methods are less subject to interference from non-specific antibody binding.

Immunocytochemical methods have been used to locate neurone-specific enolase in brain cells and in neuroendocrine cells (Schmechel *et al.*, 1978, 1978a; Tapia *et al.*, 1981).

PROPERTIES OF MULTIPLE FORMS OF ENZYMES: GENERAL CONSIDERATIONS

In spite of the very wide range of differences in properties exhibited by multiple forms of enzymes, certain broad generalizations have emerged from the numerous studies which have been made. The observation that the multiple forms of an enzyme differ in particular respects can provide indications as to their probable genetic or non-genetic origins; similarly,

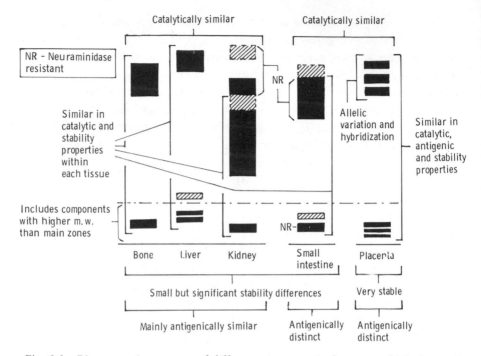

Fig. 3.6 *Diagrammatic summary of differences in properties between multiple forms of human alkaline phosphatase, some of which (e.g. between placental and non-placental phosphatases) originate at the level of the structural gene, whereas others are probably the result of post-translational modifications. The forms shown are those separated by starch-gel electrophoresis (From Moss, 1982. By permission of Karger, Basel).*

an understanding of their molecular nature can direct attention towards techniques which are likely to be effective in the analysis of mixtures of particular multiple forms. However, like most generalizations, those relating to the properties of multiple forms of enzymes are subject to numerous exceptions and qualifications. Their applicability can be seen most clearly in the case of enzymes such as human alkaline phosphatases which exist in multiple forms determined by both genetic and post-genetic factors (Fig. 3.6)

Differences in catalytic properties

The occurrence of multiple gene loci determining the structures of distinct molecules with similar, but not identical, enzymic activities and their persistence during the course of evolution presumably indicates that their existence confers advantages in natural selection; these advantages will be manifested by differences in functional properties. This argument implies that isoenzymes arising from the existence of multiple gene loci can be expected to differ to a greater or lesser extent in their catalytic properties. This important generalization has in almost every case been supported by experimental evidence, and specific differences in structure can be shown to account for functional differences in well-studied isoenzyme systems such as the carbonic anhydrases and lactate dehydrogenase. Analytical methods based on catalytic differences are therefore most likely to be useful where isoenzymes originating in this way are concerned. The advantages of such methods in the analysis of isoenzyme mixtures are that, with suitable alterations in the reaction conditions, normal quantitative methods of determining enzyme activity are used. However, the results can usually only be interpreted in quantitative terms when not more than two isoenzyme species are present in the mixture.

Multiple forms of enzymes due to post-translational modifications might be expected not to differ significantly in their catalytic properties, so that resemblance between multiple forms in this respect becomes a valuable indication that they arise by post-genetic modifications rather than at the level of the structural gene. When differences are found between non-genetic multiple forms in such characteristics as Michaelis constants they are generally slight, and analytical methods based on differences in catalytic properties are correspondingly of limited help. However, some post translational modifications of enzymes occurring within living cells do result in a significant change in catalytic properties and may therefore be of functional significance.

The differences in catalytic properties between allelozymes range from none, through minor and profound changes, to a total loss of catalytic activity by some products of rare mutant genes. This is in accordance with the expectation that the single amino acid substitutions which account for most of these isoenzymes might or might not affect the composition or conformation of the active centre.

Differences in antigenicity

The structural differences between isoenzymes that are the products of distinct gene loci typically result in differences in their antigenicity. However, allelozymes often share common antigenic properties, and the product of a mutant gene may be immunologically recognizable even when no catalytic activity remains.

As in the case of other properties, the effects of various types of post-translational modification on antigenicity are less easily summarized. In general, such processes as aggregation or association with non-protein components do not affect the antigenic properties of the enzyme protein. Although differences in carbohydrate composition are well recognized as underlying the antigenic individuality of blood group substances, variations in carbohydrate components between multiple forms of enzymes are frequently not associated with differences in antigenic specificity. This is presumably because the variable residues, e.g. sialic acid, are of such wide distribution in living matter that they do not act as haptens.

Like differences in catalytic activity, differences in reactivity with antisera offer a simple quantitative approach to the analysis of mixtures of isoenzymes determined by multiple gene loci. However, these methods also resemble those based on catalytic properties in that the results they give are more easily interpreted when only two isoenzymes are present, since hybrid isoenzymes will cross-react with antisera to either of the corresponding homopolymeric forms.

Differences in resistance to inactivation

An important property in which differences between multiple forms of enzymes are usually found is in their relative resistance to inactivation under various conditions. Differences in this respect are almost invariably found between true isoenzymes, whether these are due to the existence of multiple loci or multiple alleles. Differences between isoenzymes in their stabilities can be accounted for in terms of altered hydrophobic interactions or hydrogen bonding within the three-dimensional structure of the enzyme molecules, arising in turn from alterations in primary structure. The observation that even the single amino acid substitutions that account for the difference between allelozymes are often accompanied by changes in stability shows the great sensitivity with which stability reflects molecular structure. However, the effect of specific structural changes on stability cannot usually be predicted, nor can differences in stability be interpreted in structural terms.

Some interactions between enzymic and non-enzymic components may have a stabilizing effect, although in many cases stability is unaffected by this type of modification. Similarly, the presence or absence of some carbohydrate residues, such as sialic acid, often has no effect on stability though other properties may be markedly dependent on the presence of these strongly ionized groups.

Selective inactivation under controlled conditions can be made the basis of quantitative as well as qualitative analysis but again, measurements of the amounts of individual components in a mixture are usually only obtained when not more than two components are present unless additional components have particularly distinctive stability characteristics.

Differences in charge or size

Separative methods, especially when followed by characterization and quantitation of the individual forms, remain the definitive procedures for analysis of mixtures of multiple forms of enzymes. The observation of heterogeneity of an enzyme sample in any other respect, e.g. in stability or catalytic properties, implies the existence of different molecular forms which, at least in theory, are capable of being separated.

Differences in net molecular charge occur with equal frequency amongst genetic and non-genetic categories of multiple enzyme forms. Chemical or enzymic modification may be invoked to accentuate or to try to elucidate the structural differences underlying variations in molecular charge, but in most cases only inferences, rather than definitive conclusions, can be drawn from the results of such experiments.

Separation on the basis of differences in molecular size is unlikely to be of use when the components in question arise from differences at the genetic level, since the primary structures of isoenzymes are highly conserved, even when they are derived from loci which have become well differentiated during the course of evolution. Differences in size between them are therefore generally so small as to be beyond the resolving power of most analytical techniques. Consequently it follows that, when differences in size are found to exist between multiple forms of enzymes, these are likely to indicate the operation of post-translational processes such as proteolysis, aggregation or complex formation. As with separations based on charge differences, the effects of various modifications on the size of multiple molecular forms of enzymes can provide some information as to the possible nature of the differences between the forms.

For routine analytical purposes, separative methods have the disadvantage that they are relatively time consuming. However, against this must be set their ability potentially to resolve multi-component mixtures.

4 Distribution and Biological Functions of Multiple Forms of Enzymes

The differences in catalytic properties which are typically apparent between isoenzymes determined by multiple gene loci have presumably provided a means of adapting patterns of metabolism to the evolving needs of the organism in which they occur. It is to be expected, therefore, that the distribution of isoenzymes will not be uniform throughout the organism and wide variations in the activity of different isoenzymes do indeed occur, between organs, between the cells which compose a particular organ, and even between the structures which constitute a single cell. Equally, it should be possible to discern a correlation between the catalytic properties of individual isoenzymes and the metabolic processes of the cells or tissues in which they occur; i.e. it is to be expected that the distribution and function of isoenzymes will be closely related.

Similar arguments are less easily applied to non-isoenzymic multiple forms of enzymes, which are generally similar in their catalytic properties. Nevertheless, specific differences in distribution of these multiple forms of enzymes are also found. Whatever their evolutionary origins or functional consequences, well-established patterns of distribution account for much of the value of studies of multiple enzyme forms in biology and medicine, not only by providing insights into the metabolic patterns of different tissues and their derangements in disease, but also as the basis for organ-specific diagnosis by isoenzyme measurements.

DISTRIBUTION OF ISOENZYMES DETERMINED BY MULTIPLE LOCI

When several gene loci determine a particular type of enzyme activity, the respective isoenzymes determined by the different loci are present in each tissue in which the loci are active, together with any hybrid isoenzymes which may be formed between the different gene products. The relative proportions of the isoenzymes in a given tissue depend on their biological half-lives; i.e. on the resultant effects of differential rates of formation and breakdown of the various isoenzymes in that tissue (Fritz and Pruitt, 1977). For example, contributions of differential rates of synthesis and breakdown to the observed pattern of distribution of isoenzymes have been estimated in the case of *lactate dehydrogenase*. In the rat, isoenzyme 5 (the M_4 tetramer) is synthesized about four times faster in liver than in heart and about twice as fast in skeletal muscle. This isoenzyme is also degraded ten to twenty times more rapidly in heart, so that the half-life of LD_5 in heart is a tenth or less of that in the other two tissues (Fritz *et al.*, 1969). This implies that, in tissues in which the half-life of one gene product greatly exceeds that of the other, almost all of the lactate dehydrogenase activity will exist in the form of tetramers of that polypeptide, i.e. as either mainly isoenzyme 1 (H_4) or isoenzyme 5 (M_4). The smaller amounts of the less abundant subunit will be present as decreasing amounts of the mixed tetramers, e.g. in the order H_3M, H_2M_2 and HM_3 in tissues in which H-subunits are predominant. The homopolymer of the less favoured polypeptide (e.g. M_4) may be undetectable. When the synthetic activity of the two genes is nearly equal and their products have similar half-lives the most common isoenzyme is LD_3, the H_2M_2 tetramer, if association of the subunits into active tetramers is assumed to be random. However, this latter process may itself offer further opportunities for tissue-specific regulation of isoenzyme levels, and peptides have been identified in human liver which specifically inhibit the association of either H- or M-subunits of lactate dehydrogenase to form active tetramers (Schoenenberger *et al.*, 1980).

The operation of factors such as these helps to explain the characteristic distribution of the main lactate dehydrogenase isoenzymes in human and animal tissues, for which three basic patterns can be distinguished (Fig. 4.1). The electrophoretically more anodal isoenzymes LD_1 and LD_2 predominate in tissues such as cardiac muscle, kidney and erythrocytes, whereas in liver and skeletal muscle the more cathodal LD_4 and LD_5 isoenzymes are prominent. Isoenzymes of intermediate mobility account for the lactate dehydrogenase activity of many tissues, e.g. endocrine glands, spleen and lymph nodes, and non-gravid uterine muscle. A different, sixth lactate dehydrogenase isoenzyme, LD_X or LD_C, is present in sperm or extracts of post-pubertal testis (Fig. 6.2). This is composed of a third type of subunit which is apparently the product of a distinct

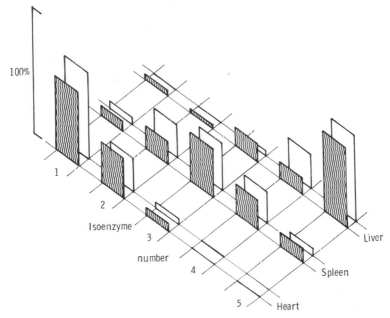

Fig. 4.1   *Distrbution of lactate dehydrogenase activity among its five isoenzymes in extracts of human heart, spleen and liver (filled bars) compared with the distributions expected for random association of the H and M monomers (open bars), assuming that the H subunit is ten times as abundant as the M subunit in heart, with these proportions reversed in liver, and that the two types of subunit are present in equal amounts in spleen. The agreement between observed and expected distributions is quite close in heart and spleen. The less good agreement in liver may be due to such factors as non-random association of subunits or the existence of a mixture of cell types with differing relative complements of the two protomers.*

structural gene expressed only in this tissue at an appropriate stage of development (Goldberg, 1977).

The distribution of lactate dehydrogenase isoenzymes within tissues has been less thoroughly studied, but some variations in isoenzyme content between different regions or cells of certain tissues have been reported. A higher proportion of LD_1 has been found in the cortical regions of rat kidney than in the medulla, although human kidney cortex and medulla appear to differ little in isoenzyme composition (Richterich *et al.*, 1961; 1963). Isoenzyme 5 is almost entirely responsible for the lactate dehydrogenase activity of purified parenchymal cells from rat liver, whereas the electrophoretically faster isoenzymes LD_4, LD_3 and LD_2 are also present in significant amounts in Küpffer cells (Berg and Blix, 1973).

Particular interest attaches to the distribution of lactate dehydrogenase isoenzymes amongst muscle fibres of different types because of the postulated differences in function between the isoenzymes, discussed later. In species in which red (or Type I) muscle fibres can be distinguished clearly from white (or Type II) fibres, the latter contain chiefly LD_5 and the former a greater proportion of the more anodal isoenzymes. A distinction between red and white skeletal muscles cannot be drawn readily in man, most human muscles containing a mixture of fibre types. However, studies on separated fibre types from human muscle suggest that the distribution of lactate dehydrogenase isoenzymes in human muscle is broadly similar to that in other vertebrates. Red fibres obtained from human skeletal muscle by microdissection more frequently showed the faster lactate dehydrogenase isoenzymes and white fibres the slower forms (Van Wijhe *et al.*, 1964). The faster isoenzymes were also more prominent in muscles in which histochemical methods indicated the presence of a high proportion of Type I fibres, but this correlation between lactate dehydrogenase isoenzyme distribution and apparent type of fibre was not constant (Rosalki, 1968).

Tissue-specific distributions of isoenzymes of many other enzymes can be accounted for by the formation of homo- and heteropolymers between polypeptide subunits which are themselves determined by differentially-active, separate structural genes. *Aldolase* resembles lactate dehydrogenase in that it is a tetrameric enzyme with subunits determined by three separate loci. As with lactate dehydrogenase, only two of the loci, those producing A and B subunits, appear to be active simultaneously in some tissues, such as kidney, giving rise to an isoenzyme pattern which consists of varying proportions of the components of a five-membered isoenzyme set. Two members of the set, the most cathodal and most anodal on electrophoresis, correspond to the A and B homopolymers, respectively. The locus which determines the structure of the C subunit is active in brain tissue, as is the A locus, so that this tissue contains aldolases A and C together with the three corresponding heteropolymers.

The existence of genetically distinct monomers of dimeric enzymes, giving rise to up to three isoenzymes, also provides the explanation for the tissue-specific distributions of such isoenzymes as alcohol dehydrogenase and creatine kinase. Three human gene loci appear to control the structures of three different subunits, α, β and γ, which can form active *alcohol dehydrogenase* dimers (Smith *et al.*, 1971a). The expression of these loci varies from tissue to tissue, as well as during the course of development. Three loci are expressed in mature liver, so that six isoenzymes are present: the $\alpha\alpha$, $\beta\beta$ and $\gamma\gamma$ homodimers, together with the $\alpha\beta$, $\alpha\gamma$ and $\beta\gamma$ hybrid isoenzymes. In other tissues only one locus is principally expressed, e.g. ADH_2 in lung and ADH_3 in the gastrointestinal tract.

The existence and differential activities of two structural genes, each determining the amino acid sequence of a specific polypeptide, also accounts for multiple forms of *creatine kinase* and their specific distributions in human and animal tissues. The electrophoretically most anodal homopolymer is present in large amounts in human brain, although it is not the exclusive form of creatine kinase in this tissue (Lindsey and Diamond, 1978), and it is consequently designated BB and its constituents as B subunits. The BB dimer is also the major form of creatine kinase present in prostate, thyroid, kidney, stomach, bladder and lung, but in smaller amounts than in brain. The more cathodal MM dimer, made up of M subunits, is the predominant enzyme form in both cardiac and skeletal muscle. There is general agreement that cardiac muscle also contains a substantial proportion of the isoenzyme with intermediate electrophoretic mobility and which consists of MB dimers, variously estimated as accounting for between 10 and 50% of the total activity of the tissue. This range of estimates is partly accounted for by inter-species differences in isoenzyme distribution; e.g. the contribution of the MB isoenzyme to the total creatine kinase activity is considerably lower in canine than in human heart. Furthermore, considerable variation between different human hearts in their MB contents has been reported (Wilhelm, 1979). The proportion of MB isoenzyme in human skeletal muscle is considerably less than in cardiac muscle, and may be undetectable by some procedures in extracts of skeletal muscle.

The MM isoenzyme of creatine kinase itself appears to be heterogeneous. Evidence has been obtained from isoelectric focusing experiments that two distinct M subunits may exist, each capable of forming active dimers with other M or B subunits (Wevers *et al.*, 1977; 1977a). This microheterogeneity of the M subunit is presumably due to some post-synthetic modification, in the way that deamidation of a single asparagine residue of aldolase A from rabbit muscle generates additional tetramers containing up to four modified subunits, which can be separated by isoelectric focusing (Horecker, 1975).

Numerous examples of tissue-specific distributions of isoenzymes resulting from the differential expression of multiple gene loci can be added to those already cited. Among enzymes of carbohydrate metabolism, *enolase* resembles aldolase in being determined by three genetic loci which are differentially expressed in human and rat tissues (Pearce *et al.*, 1976; Rider and Taylor, 1974, 1975, 1976). Since enolase is also a dimeric enzyme, the three distinct polypeptides which are encoded by these loci can potentially give rise to six homodimeric and heterodimeric iso-enzymes. Five of the possible isoenzymes have been detected in human tissues in relative amounts which vary from one tissue to another. Skeletal muscle contains almost exclusively the homodimeric product of a single locus, whereas a different locus is expressed predominantly in liver and other tissues. Two loci are expressed to significant degrees in brain; one is the locus which is also expressed in tissues such as liver, but the second locus is expressed almost exclusively in nervous tissue. This tissue contains a hybrid enolase, as well as two homodimeric forms. The brain-specific enolase isoenzyme occurs in neurones in rat brain, whereas the non-specific isoenzyme is confined to non-neuronal cells such as glial cells (Schmechel et al., 1978). Neurone-specific enolase has also been located in central and peripheral neuroendocrine cells of the APUD (amine pre-cursor, uptake and decarboxylation) system (Schmechel et al., 1978a) and in neuroendocrine tumours (Tapia et al., 1981).

Four isoenzymes of *pyruvate kinase* have been distinguished in vertebrate tissues (Hall and Cottam, 1978). The M_1 isoenzyme contributes virtually the whole activity of this enzyme in adult skeletal muscle, and most of it in cardiac muscle and brain.* The L-type pyruvate kinase is the predominant form in liver, and has markedly different properties from the M_1 isoenzyme. Pyruvate kinase is confined to parenchymal cells, but a minor component, M_2, is also present in the non-parenchymal elements of the liver (Van Berkel et al., 1972). Pyruvate kinase M_2 is widely distributed and is the main isoenzyme in kidney and leucocytes. The fourth pyruvate kinase isoenzyme, R, is the only form of the enzyme found in red blood cells. The four isoenzymes can be separated electrophoretically, their anodal mobilities decreasing in the order L, R, and, of nearly equal mobilities, M_2 and M_1.

Kinetically, the pyruvate kinase isoenzymes fall into two groups: the M_1 isoenzyme exhibits Michaelis–Menten kinetics, whereas the M_2, L and R forms are allosteric enzymes. However, there are quantitative kinetic differences between the latter forms. Structurally, all the isoenzymes are tetramers; these are probably homopolymers of four characteristic subunits in the case of each isoenzyme except the R form, for which evidence has been obtained for the presence of two types of subunits (Peterson et al., 1974). Structural similarity between the M_1 and L isoenzymes has been demonstrated by hybridization *in vitro* (Cardenas

---

*Many different designations of the isoenzymes of pyruvate kinase have been used (cf. Ibsen, 1977).

and Dyson, 1973). *In vivo* hybridization of the M_2 and L isoenzymes has been confirmed by the generation of additional isoenzymes from a presumed $(M_2)_2L_2$ band isolated from bovine kidney (Cardenas *et al.*, 1975), and what appear to be M_2–L hybrids have been observed in human hepatomata and their host livers (Hammond and Balinsky, 1978).

Immunological studies show cross-reactivities between the M_1 and M_2 isoenzymes on the one hand, and isoenzymes L and R on the other (Hall and Cottam, 1978). Pyruvate kinase synthesized in red blood cells as subunits of molecular weight 63 000 can be converted *in vitro* to a form similar in molecular weight to that characteristic of liver tissue, by limited tryptic attack (Kahn *et al.*, 1978). The transformation is accompanied by an enhancement of regulatory properties. These observations suggested post-genetic processing, specific to liver, of a precursor common to both reticulocytes and liver cells. However, although genetic evidence strongly supports the existence of a single structural gene, by which both liver-type and erythrocyte-type pyruvate kinase subunits are coded, the messenger RNA molecules produced by the gene are not identical when it is expressed in the two cell types. Therefore, instead of selective post-genetic modification of a common polypeptide, differential processing of a common nuclear RNA precursor, or rearrangement of the gene in one type of cell during differentiation have been postulated (Marie *et al.*, 1981).

Multi-locus determination of the structures of distinct monomers also accounts for the occurrence of isoenzymes of *phosphofructokinase* with tissue-specific distributions in man and animals (Tsai *et al.*, 1975; Vora *et al.*, 1980). The predominant isoenzymes of liver and skeletal muscle, with distinct catalytic and physical properties, are respectively the homopolymers of two types of subunits. Since the active enzyme is a tetramer, up to three additional hybrid isoenzymes can be formed.

Hexokinase exists in multiple forms with characteristic distributions in mammalian tissues. However, since this enzyme is monomeric, hybrid isoenzymes cannot exist. Three isoenzymes, usually designated I, II and III in order of increasing anodal electrophoretic mobility, are widely present in varying proportions (Purich *et al.*, 1973). In addition, a fourth enzyme with hexokinase activity, but with a much higher specificity for glucose than the others and differing from them in various respects, is almost entirely confined to the liver of those species in which its presence has been demonstrated. The classification of this enzyme, glucokinase, as a hexokinase isoenzyme is not universally accepted. Hexokinase isoenzyme I, the electrophoretically least anodal form, is present in high activity in brain, liver, kidney and lung. Like isoenzyme III, which is also present in significant amounts of liver, kidney and lung, the activity of isoenzyme I is not increased by administration of insulin. Type II hexokinase is present in skeletal and cardiac muscles and liver; as with glucokinase, with which it occurs in liver in some species, the activity of this isoenzyme is influenced by insulin.

The pancreatic and salivary isoenzymes of human *amylase* are respectively found mainly in those tissues, as their names indicate. These isoenzymes are the products of separate though closely-linked loci on chromosome 1, but the gene products are subjected to various post-translational modifications (Karn *et al.*, 1974). Other genes also are expressed exclusively, or almost so, in a single tissue, in some cases at a particular stage in the development of the organism. An example which has already been mentioned is the third lactate dehydrogenase locus, active in mature testis, which determines the subunit composing the third homopolymer of the enzyme, LD_X (Goldberg, 1977). The isoenzyme of alkaline phosphatase which occurs in the human placenta is the product of a single structural gene locus, distinct from the two or possibly more additional loci which specify the structures of other forms of alkaline phosphatase, and the product of the placental-phosphatase locus is normally detectable only in placenta (Donald and Robson, 1974). The distinctive 'prostatic' isoenzyme of acid phosphatase is produced in characteristically large amounts in that tissue and differs in its properties from the various more widely-distributed acid phosphatases.

Few generalizations can be made about the tissue-specific patterns of expression of multiple isoenzyme-determining loci. In general, qualitatively similar distributions of isoenzymes are found in homologous tissues from different species, although quantitative differences occur. For example there are large variations from one species to another in the relative proportions of the H_4 and M_4 isoenzymes of lactate dehydrogenase in erythrocytes and liver. Certain tissues can be distinguished from others by their possession of specific variants of each of several enzymes which form part of a particular common metabolic pathway; e.g. many of the glycolytic enzymes exhibit 'liver-' or 'muscle-type' isoenzymes. There is a tendency for the more acidic isoenzymes of some enzymes, such as creatine kinase, aldolase and enolase, to be present in brain. However, hexokinase provides an exception to this, in that it is the more basic isoenzyme I that predominates in this tissue.

The distribution of isoenzymes among the tissues has so far been almost entirely studied by the determination of their specific catalytic activities. However, the availability of specific antisera to several isoenzymes or their constituent subunits increasingly provides opportunities to detect inactive forms, and thus to reassess the contribution of factors such as gene expression or selective isoenzyme inactivation to the observed patterns of activity. For example, the presence of large amounts of immunoreactive but catalytically-inactive BB creatine kinase has been demonstrated in mature rabbit and human skeletal muscle and in rabbit cardiac muscle (Armstrong *et al.*, 1977).

Intracellular distribution of isoenzymes

The isoenzyme systems so far discussed are largely confined to the same compartment of the cells in which they occur; in most cases this is the cytoplasm, although hexokinase isoenzymes are known to bind select-ively to subcellular structures (Katzen and Soderman, 1975). However, many instances are known in which the expression of different structural genes gives rise to characteristic distributions of isoenzymes amongst the various cellular organelles (Holmes and Masters, 1979).

Mitochondrial isoenzymes

Several isoenzymes located in mitochondria are different from their counterparts in other regions of the cell. Well-recognized examples include mitochondrial *malate dehydrogenase* and *aspartate aminotrans-ferase*. The structures and properties of the cytoplasmic and mitochon-drial variants of these enzymes have been extensively studied and compared in several species. The independent genetic origins of the isoenzymes of malate dehydrogenase have been confirmed in mouse and man by the discovery of rare variants of the mitochondrial isoenzyme which are inherited in a Mendelian fashion without corresponding changes in the cytoplasmic form (Davidson and Cortner, 1967; Shows *et al.*, 1970). The gene determining mitochondrial malate dehydrogenase has been mapped to chromosome 7 and that determining the cytoplasmic isoenzyme to chromosome 3 in man (Shows, 1977). The existence of separate genes controlling the structures of cytoplasmic and mitochon-drial aspartate aminotransferase isoenzymes has similarly been deduced from observations of rare mutants in man and mouse, with assignments of the mitochondrial and cytoplasmic isoenzyme genes to mouse chromo-somes 8 and 19, respectively (Davidson *et al.*, 1970; Chen and Giblett, 1971; Delorenzo and Ruddle, 1970).

Among other isoenzymes which exist in distinct mitochondrial and cytoplasmic forms are the malic enzyme and NADP-dependent isocitrate dehydrogenase, superoxide dismutase (Fridovich, 1975), aconitase (Slaughter *et al.*, 1977) and adenylate, creatine and thymidine kinases (Berk and Clayton, 1973).

Genetic variation of the *malic enzyme* of the mouse indicates a tetrameric structure (Shows *et al.*, 1970) and the genes which code for the mitochondrial and cytoplasmic isoenzymes have been mapped to chromosomes 7 and 9, respectively, in this species. Separate gene loci for mitochondrial and cytoplasmic *isocitrate dehydrogenase* have also been confirmed in the mouse and in man, the genes having been assigned respectively to human chromosomes 15 and 2 (Shows, 1977). The gene determining the cytoplasmic isoenzyme is duplicated in the rainbow trout, giving rise to a tissue-specific distribution of the two cytoplasmic isocitrate dehydrogenases in this species (Reinitz, 1977).

As is the case for other isoenzymes determined by distinct genetic loci, the mitochondrial and cytoplasmic isocitrate dehydrogenases from mammalian tissues differ in physicochemical properties such as electrophoretic mobility, stability to heat and immunogenicity, as well as having quantitatively different catalytic characteristics (Campbell and Moss, 1962; Lowenstein and Smith, 1962; Islam *et al.*, 1972; Carlier and Pantaloni, 1973). Further multiple forms, or 'secondary isoenzymes', of isocitrate dehydrogenase have been described as they have for the isoenzymes of malate dehydrogenase (Turner *et al.*, 1974).

The cytoplasmic isoenzyme (designated 'A' or '1') of *adenylate kinase* is present in high activity in mammalian muscle, brain and erythrocytes and is determined in man by a gene located on chromosome 9. It is distinct from isoenzyme B (or 2) found in the mitochondria of liver and other tissues (Itakura *et al.*, 1978), the gene for which is carried on chromosome 1 in man (Shows, 1977). A third 'isoenzyme' of adenylate kinase, also located in mitochondria, with a wider specificity for the triphosphate nucleoside substrate of the catalysed reaction, may more properly be regarded as a separate enzyme and has been given an individual number (EC 2.7.4.10) in the Enzyme Commission's list.

Creatine kinase is generally regarded as essentially a soluble enzyme, although a small proportion of the MM isoenzyme is associated with the M-line region of skeletal muscle of the chicken, while some of the BB creatine kinase which predominates in chicken cardiac muscle is tightly bound to the Z-line fraction (Eppenberger *et al.*, 1975). Creatine kinase MM has been reported to be associated with the surface membrane of myocardial cells (Saks *et al.*, 1977). However, a mitochondrial creatine kinase seems also to exist which, because of its different catalytic properties, probably represents a third isoenzyme (Jacobs *et al.*, 1964). This enzyme form migrates just cathodally of the MM isoenzyme on electrophoresis. The mitochondrial isoenzyme purified from beef heart is a dimer of 64 000 molecular weight, which does not form hybrid isoenzymes *in vitro* with other creatine kinase isoenzymes from this tissue (Hall *et al.*, 1979). Antiserum to human M-type creatine kinase subunits did not reduce the creatine kinase activity of a mitochondrial preparation from human brain (Lindsey and Diamond, 1978). The mitochondrial creatine kinase may therefore be assumed to be structurally distinct from other isoenzymes. Its catalytic properties are also different: maximum velocities for the forward and reverse reactions catalysed by mitochondrial creatine kinase are approximately equal, but differ by a factor of four in the presence of the MM isoenzyme. Mitochondrial creatine kinase appears to be located on the outer surface of the inner mitochondrial membrane (Jacobus and Lehninger, 1973).

The contribution of mitochondrial isoenzymes to the total specific enzyme activities of different tissues will in many cases reflect the relative abundance of mitochondria in them; e.g. a high concentration of these isoenzymes is to be expected in mitochondria-rich organs such as liver, in which the mitochondrial isoenzyme accounts for more than half of the total aspartate aminotransferase activity of parenchymal cells. However, differences do exist between mitochondria from different tissues with respect to their content of particular isoenzymes. Adult human heart, also a tissue which is rich in mitochondria, contains less cytoplasmic than mitochondrial isocitrate dehydrogenase, whereas in liver, the relative proportions of the two isoenzymes are reversed (Campbell and Moss, 1962). The mitochondrial isoenzyme of aconitase is also less prominent, relative to the cytoplasmic isoenzyme, in human liver than it is in heart (Slaughter *et al.*, 1977). Mitochondria from human, bovine and rat hearts, and also from brain and striated and smooth muscle, contain considerable creatine kinase activity but virtually no activity is detectable in mitochondria from rat or rabbit liver, kidney or testis (Jacobus and Lehninger, 1973).

The possible biochemical differences between mitochondria from different tissues are not fully clarified, although morphological differences are well recognized. However, since chromosomal mapping has confirmed that the structures of many mitochondrial enzymes are determined by nuclear genes, there seems to be no reason why these genes should not be expressed to different extents in different tissues, as are the genes controlling cytoplasmic isoenzymes, to produce tissue-specific distributions of mitochondrial as well as cytoplasmic isoenzymes

Although in almost every case independent mutation or chromosome mapping has confirmed the isoenzymic nature of analogous mitochondrial and extra-mitochondrial enzymes, such studies point to a single genetic origin for the cytoplasmic and mitochondrial forms of *fumarase* in human tissues. Electrophoresis of tissue extracts fractionates this enzyme into two sets of zones, one set associated with cytosol and the other, less anodal, with the mitochondria (Tolley and Craig, 1975; Edwards and Hopkinson, 1979). A tetrameric structure has been deduced for the corresponding enzyme of rat liver, from observations of a five-membered isoenzyme set in rats which are heterozygous for variant forms of fumarase, and by generation of corresponding zones in hybridization experiments *in vitro* with isoenzymes from rats homozygous for the different genetic variants (Carleer and Ansay, 1976). However, the heterogeneity of the common forms of both mitochondrial and cytoplasmic fumarases in man seems to be due to post-genetic modification.

The evidence for a single gene locus determining both forms comes from the observation that a rare mutation alters the electrophoretic patterns of the human mitochondrial and cytoplasmic fumarases in a similar way. In an individual presumed to be heterozygous for the normal and variant alleles, both the soluble and mitochondrial fumarase components included additional bands consistent with the formation of hybrid tetramers between normal and modified subunits (Edwards and Hopkinson, 1979; 1979a).

These observations on human fumarase, though probably representative of only a small group of enzymes or possibly even constituting a single exceptional case, nevertheless preclude a generalized statement that analogous cytoplasmic and mitochondrial enzymes are always the products of distinct structural genes.

Lysosomal isoenzymes

Lysosomes contain a range of hydrolases, some of which exist in multiple forms within these organelles, while many have isoenzyme counterparts elsewhere in the cell. Often the lysosomal isoenzymes exhibit the more acid pH optima, compared with extra-lysosomal forms, which are typical of lysosomal hydrolases in general. The separate genetic origins and thus the isoenzymic status of the lysosomal enzymes has been indicated in several instances by the recognition of inherited storage diseases in which particular lysosomal enzymes are deficient (Chapter 5).

Three glycosidases optimally active at acid pH occurring in lysosomes from mammalian cells can be distinguished from analogous cytoplasmic isoenzymes. Lysosomal *β-D-glucosidase* is a tetramer of 240 000 molecular weight, specific for the hydrolysis of β-D-glucopyranoside linkages (Pentchev *et al.*, 1973; Kanfer *et al.*, 1975). Its absence from human lysosomes is the cause of Gaucher's disease, in which its natural glucocerebroside substrate accumulates. The soluble isoenzyme of β-glucosidase isolated from rat kidney is a monomer of 50–58 000 molecular weight and with a somewhat more alkaline pH optimum of 5.5–7.0 and a relatively low substrate specificity (Glew *et al.*, 1976). However, although activity of the human cytoplasmic enzyme is not reduced in Gaucher's disease, it cannot hydrolyse glucocerebroside.

A deficiency of human lysosomal *β-D-galactosidase* results in the storage disease of generalized gangliosidosis, in which GM_{1-} ganglioside accumulates. The isoenzyme from human liver has a molecular weight of 60–70 000 (Norden *et al.*, 1974). Cytoplasmic β-galactosidase, with a less acidic pH optimum, is not reduced in activity in patients with this disease. The lysosomal and cytoplasmic isoenzymes are immunologically distinct (Meisler and Rattazzi, 1974), but the separate identities of cytoplasmic β-galactosidase and β-glucosidase have not been finally established (Meisler, 1975).

Immunochemically and biochemically distinct lysosomal and cytoplasmic *α-mannosidases* have been detected in the livers of several species (Phillips *et al.*, 1975;. Snaith, 1977). A genetically-determined selective absence of the lysosomal isoenzyme is the cause of the rare storage disease mannosidosis, in which mannose-rich oligosaccharides accumulate (Carroll *et al.*, 1972). Lysosomal α-mannosidase itself exists in multiple forms which apparently are due to the presence of varying amounts of *N*-acetylneuraminic acid since they can be modified by treatment with neuraminidase (Chester *et al.*, 1975). A second extra-lysosomal α-mannosidase appears to be associated with the Golgi apparatus (Dewald and Touster, 1973).

The isoenzymes of the lysosomal enzyme N-*acetyl-β-hexosaminidase* have been the subjects of extensive investigations because of the storage diseases, Tay-Sachs and Sandhoff's diseases, which their deficiency causes. Three lysosomal variants of hexosaminidase can be separated from all normal tissues, except red blood cells: in order of increasing anodal mobility on electrophoresis these are designated B, A, and a minor component, S. As well as differing in electrophoretic mobility, the A and B isoenzymes differ in heat stability and kinetic properties (Robinson and Stirling, 1968; Srivastava *et al.*, 1974). However, they are immunologically cross-reactive. The relationships between the isoenzymes are accounted for by their oligomeric structures, which are composed of two types of subunits, α and β. The B isoenzyme is considered to be a homopolymer of β subunits and the A isoenzyme an $\alpha\beta$ heteropolymer. Isoenzyme S is composed of α subunits only. These postulated quaternary structures have been substantiated by *in vitro* hybridization experiments (Beutler and Kuhl, 1975). Estimates of the molecular weights of the α and β subunits and of the native A and B isoenzymes, of 25–27 000 and 110 000, respectively, suggest that the isoenzymes are tetramers (Lee and Yoshida, 1976; Beutler *et al.*, 1976).

A cytosolic isoenzyme, hexosaminidase C, with rather similar mobility to the S isoenzyme occurs widely in human tissues; however, it is distinct from the A, B and S isoenzymes both kinetically and immunologically, and has a rather higher molecular weight (Poenaru and Dreyfus, 1973; Penton *et al.*, 1975). In view of these differences and the lack of alteration in its activity in Tay-Sachs or Sandhoff's diseases, it is probable that hexosaminidase C has an independent genetic origin. Minor forms of hexosaminidase, I and S', have been observed on electrophoresis, which may arise by variations in the carbohydrate moieties of the B and S isoenzymes (Beutler and Kuhl, 1975).

Non-specific *acid phosphatase* from various mammalian tissues displays very marked heterogeneity on zone electrophoresis (Fig. 4.2). Part of this heterogeneity derives from the existence of lysosomal and cytoplasmic isoenzymes, part from the existence of other multiple loci or the occurrence of common alleles at one or more loci, and part from post-genetic modifications of structure occurring either *in vivo* or *in vitro*. However, the origins and relationships of many of the multiple forms of this enzyme remain unexplained.

Low molecular weight acid phosphatase has been identified in red blood cells and in the cytoplasm of most other tissues, and appears to be a monomer of about 16 500 molecular weight. The independent genetic origin of the red cell isoenzyme, as it is usually called, is demonstrated by the occurrence of allelic variants in man, some with high frequencies, and its specific locus has been mapped to human chromosome 2 (Hopkinson *et al.*, 1963; Ferguson-Smith *et al.*, 1973). The red cells of heterozygous individuals have isoenzyme patterns which represent the sum of the zones determined by each allele, since the enzyme is a monomer.

The common allelozymes have similar kinetic properties, but differ in thermostability. Multiple electrophoretic forms are seen, even in individuals who are homozygous at the controlling locus: these forms

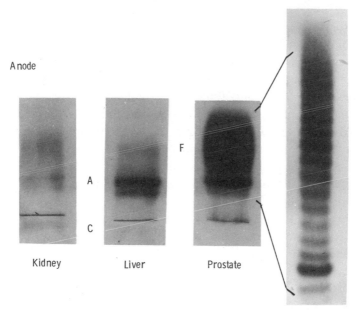

Fig. 4.2   *Acid phosphatase zones in extracts of human kidney, liver and prostate, separated by starch-gel electrophoresis at pH 6.2. Electrophoretically similar zones (A) are present in all three extracts, and a cathodal zone (C) is detectable in the kidney extract. The intense anodal zone (F) in extracts of prostate can be resolved into multiple discrete zones (right) which appear to contain different numbers of sialic acid residues.*

appear to be post-genetic modifications of the gene product. The acid phosphatases determined by the red cell phosphatase locus are distinct from other acid phosphatases in various physical and catalytic properties, e.g. in relative rates of hydrolysis of various substrates and response to inhibitors.

Lysosomal acid phosphatases from several tissues have molecular weights of the order of 100–120 000 and a dimeric structure. Their substrate specificity appears to be somewhat broader than that of the cytoplasmic isoenzymes. The number and expression of loci determining these isoenzymes is not yet certain. Two loci appear to account for three forms of the enzyme in human placenta, two forms being homodimers of the products of the respective loci and the third consisting of a hybrid dimer of the two distinct subunits. This explanation is supported by the alterations of acid phosphatase patterns in this tissue which have been observed as a result of rare allelic mutations (Beckman *et al.*, 1970; Swallow and Harris, 1972).

Some electrophoretic evidence suggests that one or both of these loci are expressed in tissues other than placenta. A zone corresponding to the more anodal ('A') placental isoenzyme is detectable, for example, in extracts of kidney, liver, spleen and lung, and the slower placental ('C') isoenzyme is found in these tissues as well as skeletal muscle and brain. Furthermore, a rare, apparently inherited, generalized deficiency of lysosomal acid phosphatase has also been described which appears to derive from changes at one or both of the lysosomal acid phosphatase loci (Nadler and Egan, 1970). However, enzyme forms which are electrophoretically faster than the A isoenzyme are present in extracts of some tissues; e.g. prostate, in which they are the most prominent components. The faster components often appear as broad, diffuse zones but they can be resolved into multiple, discrete bands (Sur *et al.*, 1962). This heterogeneity is at least partly due to the presence of different amounts of sialic acid in the zones since it is reduced by incubation with neuraminidase (Smith and Whitby, 1968).

The acid phosphatase secreted in large quantities by the human prostate gland and by prostates of some other species has become the most extensively studied of all the isoenzymes of acid phosphatase, because of its clinical importance. It has been extensively purified and shown to be a dimer of molecular weight about 100 000 at pH 5–6, which readily undergoes aggregation or dissociation (Lucher-Wasyl and Ostrowski, 1974) and its amino acid composition has been determined (Vikho, 1978). Prostatic acid phosphatase is clearly distinct from the red-cell isoenzymes in substrate specificity and inhibition characteristics, although it resembles high molecular-weight isoenzymes from some other tissues in these respects. Similarities between prostatic and non-prostatic acid phosphatases in electrophoretic behaviour have been noted, as well as quantitative and qualitative differences (Lam *et al.*, 1973; Harris and Hopkinson, 1976).

Opportunities to re-examine the relationships between prostatic and non-prostatic acid phosphatases have been provided by the development of methods of purification by affinity chromatography which exploit the inhibition of prostatic and some other acid phosphatases by L(+)-tartrate (Vikho *et al.*, 1978; Van Etten and Saini, 1978), and by the availability of antisera raised against the purified prostatic isoenzyme. Such antisera do not cross-react with acid phosphatases in erythrocytes, spleen, liver, kidney, intestine, bladder or bone (Chu *et al.*, 1978; Vikho *et al.*, 1978) although a slight reaction was observed with acid phosphatase in brain (Foti *et al.*, 1975). These observations, together with the dependence of prostatic acid phosphatase secretion on male sex hormones, suggest that this isoenzyme may be the product of a separate gene which is expressed almost exclusively in the prostate gland. Further support for this hypothesis is offered by the appearance of an acid phosphatase immunologically identical with the prostatic isoenzyme in a pancreatic islet-cell carcinoma (Choe *et al.*, 1978), perhaps signalling the derepression of a specific gene.

Acid phosphatase activity appearing in zones 1 to 4 in polyacrylamide-gel electrophoresis is inhibited by L(+)-tartrate, whereas the activity of zone 5 is not. This zone occurs in extracts of spleen from patients with leukaemic reticuloendotheliosis (Yam, 1974) and has been detected also in sera from patients with bone diseases of various kinds, as well as sera from healthy children and adults (Lam *et al.*, 1978). Tartrate-resistant acid phosphatase has been recognized as a minor component of serum for many years, and the increase in this activity often seen in serum in bone disease has been attributed to increased activity of the osteoclasts, since these cells contain a similar acid phosphatase. The tartrate-resistant enzyme also differs from the inhibitable acid phosphatases in its relative rate of hydrolysis of various orthophosphates. According to the hypothesis that true isoenzymes tend to have distinctive kinetic properties (Chapter 3), 'band 5' tartrate-resistant acid phosphatase may be supposed to be the product of a further gene locus with its expression restricted to a few tissues.

Multiple forms of enzymes in cell membranes

Several enzymes are characteristically and universally associated with the external and internal membranes of cells; others occur in specialized regions of the plasma membranes of particular cells, such as the brush borders of renal or small-intestinal epithelial cells. The possibility therefore arises that enzymes thus specifically localized are isoenzymically different from their counterparts in other regions of the cell, or that tissue-specific variants of membrane-bound enzymes may exist. Furthermore, preparations of membrane-bound enzymes are typically heterogeneous, posing additional problems of interpretation, not all of which can be solved fully at present.

Alkaline phosphatases are widely-distributed components of the plasma membranes of mammalian cells, activities being particularly high in the epithelia of the renal tubules and small intestine. Osteoblasts are also rich in the enzyme and in these cells also it is membrane-bound, while in the liver it is located in the microvilli of the canalicular surfaces of parenchymal cells. Alkaline phosphatases from these sources differ in physicochemical properties such as electrophoretic mobility, resistance to denaturation, and, in the case of the intestinal form, in the absence of terminal sialic acid residues. Differences in catalytic properties are slight, but are most marked between intestinal and non-intestinal alkaline phosphatases, taking the form of small differences in relative rates of hydrolysis of alternative substrates (Fig. 3.1) and a greater sensitivity on the part of the intestinal enzyme to inhibition by L-phenylalanine and some other compounds. These differences may indicate a separate genetic determination of intestinal alkaline phosphatase, a view which is supported by the observation that normal or even increased activities of this isoenzyme are present in the intestines of patients with the hereditary condition of hypophosphatasia, in which the levels of alkaline phosphatase in bone, and also in other non-intestinal tissues, are greatly reduced (Brydon *et al.*, 1975). Mono-specific antisera against human intestinal alkaline phosphatase have also been prepared (Lehmann, 1975; 1975a).

The alkaline phosphatase produced in large quantities in the brush border of the placental syncitiotrophoblast of man and a few other primates shares the catalytic characteristics of intestinal alkaline phosphatase, but is distinguished from this and all other alkaline phosphatases by its remarkable stability to heat. The independent genetic origin of this isoenzyme is demonstrated by its numerous inherited variations (Fig. 4.3), which are not accompanied by corresponding changes in the alkaline phosphatases of other tissues (Donald and Robson, 1974). Such evidence as is available from radioactive peptide mapping (Fig. 2.4) and comparisons of labelled placental and non-placental phosphatases also confirms the distinct primary structure of the placental isoenzyme (Whitaker and Moss, 1979; McKenna *et al.*, 1979). The patterns of placental alkaline phosphatase zones on electrophoresis of extracts of placentae from heterozygous individuals indicate that the active enzyme is a dimer; this is probably true also of non-placental phosphatases, but dissociation and hybridization *in vitro* has not been effected.

At least three gene loci, therefore, appear to specify different isoenzymes of alkaline phosphatase in man: a placental locus, an intestinal locus, and one or more loci which determine the phosphatases of other tissues (Moss, 1970). Differences in properties between non-placental, non-intestinal alkaline phosphatases are slight and probably result from post-translational modifications of a single gene product, although they do include variations in stability which may reflect underlying dif-

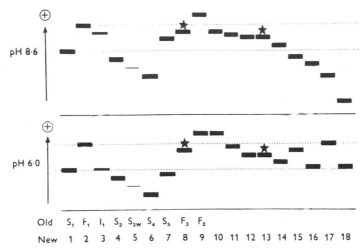

Fig. 4.3 *Diagrams of the mobilities of 18 allelozymes of human placental alkaline phosphatase on starch-gel electrophoresis at two pH values. The forms marked with asterisks can be distinguished from electrophoretically-similar variants by their differentially altered mobilities after treatment with neuraminidase. Only the major zones present in each extract are represented, with their original and revised designations (From Donald and Robson, 1974. By permission of Cambridge University Press).*

ferences in primary structures between these forms. Inferences as to the possible genetic independence of the various tissue-specific forms of alkaline phosphatase can also be drawn from their patterns of occurrence in various tissues. In general, a single, characteristic form of the enzyme predominates in each tissue.

When minor alkaline phosphatase components are detectable which are not of the type characteristic of that tissue, these are of a readily distinguishable and probably genetically distinct character. For example, extracts of kidney contain minor components of alkaline phosphatase which are unaffected in net molecular charge by treatment with neuraminidase, a property possessed by intestinal phosphatase, and which cross-react with antiserum to intestinal phosphatase (Butterworth and Moss, 1966; Boyer, 1963). Conversely, a small fraction of intestinal alkaline phosphatase is neuraminidase-sensitive (Moss *et al.*, 1966). However, bone phosphatase-like components cannot be reliably demonstrated in liver, for example, or kidney phosphatase in bone. Although these observations may be due to the difficulties of positively identifying the components of mixtures of such closely similar enzyme forms as the alkaline phosphatases of bone, liver and kidney, data on the distribution of alkaline phosphatases are consistent with the existence of distinct structural genes for the placental and intestinal isoenzymes, with minor tissue-specific modifications of the product of a third locus accounting for the alkaline phosphatases predominant in other tissues (Fig. 3.6).

The alkaline phosphatase activity in all tissue extracts exhibits some degree of heterogeneity when separated by charge-dependent means such as electrophoresis. This heterogeneity is most marked in extracts of kidney or small intestine, in which very diffuse main zones of activity are seen, and much of it can be accounted for by the presence of varying numbers of terminal sialic acid residues in kidney phosphatase and to the non-covalent attachment of lipids and low molecular weight peptides to the intestinal enzyme (Butterworth and Moss, 1966; Nayudu and Hercus, 1974). As well as the heterogeneous main zones of alkaline phosphatase in tissue extracts, minor electrophoretic zones are often present. Since these resemble the major active fractions of the same extract in such organ-specific properties as stability and antigenicity, their presence can be attributed to aggregation or complex-formation of a single enzyme type in each tissue (Moss and King, 1962).

Other membrane-bound enzymes display a similar degree of hetero-geneity to that shown by alkaline phosphatase, with the effect that the possible existence of tissue-specific variants may be obscured. This is the case, for example, for the enzyme *γ-glutamyl transferase* which occurs in high concentrations in organs such as pancreas and kidney, and which is of clinical interest because of the elevation of its activity in serum in hepatobiliary disease. Preparations of the enzyme from tissues or serum are markedly heterogeneous with respect to both size and molecular charge (Fig. 4.4); however, there are no significant differences in catalytic characteristics between the various forms (Echetebu and Moss, 1982). The native enzyme appears readily to undergo interactions such as aggreg-ation or complex-formation which change its physical properties. Various experimental treatments, including proteolysis, extraction with organic solvents or detergents, or treatment with neuraminidase, modify the size and charge of the multiple enzyme forms (Echetebu and Moss, 1982a). The multiple forms of γ-glutamyl transferase thus appear to reflect the various interactions and associations of the enzyme molecule, as a consequence of its incorporation into the structure of membranes in which it occurs. γ-Glutamyl transferase treated with proteolytic enzymes loses its interactive characteristics and consists of two subunits of unequal size, the smaller of which carries the active centre (Curthoys and Hughey, 1979).

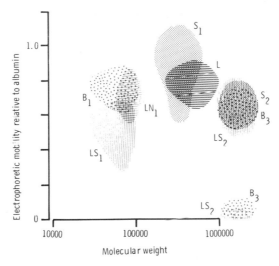

Fig. 4.4   *Heterogeneity of γ-glutamyl transferase in human serum, bile and extracts of liver tissue with respect to net molecular charge and size. The fractions are designated as follows: S_1, S_2, prepared from serum; B_1, B_3, prepared from bile; L, extracted from liver by detergent solutions; LS_1, LS_2, prepared by incubating liver tissue in serum* in vitro; LN_1, *prepared by incubating liver tissue in sodium chloride solution* in vitro *(From Echetebu and Moss, 1982. By permission of Karger, Basel).*

The heterogeneity introduced into enzyme preparations by their incorporation into membranes has been more extensively studied in the case of the *β-glucuronidase* of mouse tissues (Lusis and Paigen, 1977). This enzyme is located both in lysosomes and in the internal membrane system, the endoplasmic reticulum, of many tissues. Both forms have similar catalytic properties, and both are tetramers composed of subunits of about 70 000 –75 000 molecular weight determined by a single structural gene. However, the subunits incorporated into the lysosomal and microsomal tetramers are not identical, the subunits of the lysosomal enzyme being somewhat reduced in size and positive charge by post-translational modifications. The larger protomers associate to form the microsomal enzyme, but microsomal β-glucuronidase can be separated into four forms with differing molecular weights. These additional forms arise from the association of the enzyme tetramers with one, two, three or four molecules of a non-enzymic protein which has been called 'egasyn'. Egasyn is the product of a distinct gene locus, as is demonstrated by a genetic deficiency of its production in mice of the mutant YBR strain which consequently lack microsomal, but not lysosomal, β-glucuronidase. The egasyn-glucuronidase complexes can be released intact from the microsomes;

however, more vigorous treatments, e.g. with concentrated urea solutions, break the association between the enzyme and egasyn.

The post-translational conversion of the slightly larger glucuronidase tetramers to the form characteristic of the lysosomal enzyme, which appears to be enzyme-catalysed, presumably prevents the interaction of the modified glucuronidase with egasyn. Thus, the level of activity of this modifying system may provide a means of regulating the relative proportions of the lysosomal and microsomal enzymes. The distribution of the two forms of glucuronidase can also be affected by the availability of the binding protein egasyn; treatment with phenobarbitone stimulates production of microsomal proteins, including egasyn, in mouse liver and the ratio of microsomal to lysosomal glucuronidase increases, although the total glucuronidase activity of the liver remains constant.

Effect of multiple allelism on isoenzyme distribution

The occurrence of allelic genes at isoenzyme-determining loci may modify specific patterns of isoenzyme distribution. This is seen most clearly when an allele has no recognizable product, resulting in the total disappearance of an isoenzyme from its characteristic locations in homozygous individuals. Where the product of the allelic gene is a variant isoenzyme, heterozygous individuals exhibit additional isoenzyme patterns superimposed on the distributions seen in homozygotes for the common allele.

PHYSIOLOGICAL FUNCTION OF MULTIPLE FORMS OF ENZYMES

The differences which typically exist between isoenzymes determined by multiple gene loci in such characteristics as dependence of activity on substrate concentration or response to activators or inhibitors provide a basis for attempts to explain their physiological functions in terms of these differences. Such attempts have carried varying degrees of conviction. An early success of this approach was the demonstration that three isoenzymes of aspartate kinase in *E. coli* are respectively inhibited by lysine, methionine and threonine. Since each of the synthetic pathways leading to these three amino acids begins with the phosphorylation of aspartate, the existence of the three isoenzymes allows feedback inhibition of its own production by each amino acid to take place, without affecting production of either of the other two (Stadtman, 1968). Several similar examples of the regulation of branched metabolic pathways in bacteria through the existence of isoenzymes have also been demonstrated.

Compartmentalization of metabolic pathways in tissues and organelles

The existence of multiple structural gene loci determining functionally similar, though not identical, isoenzymes provides a means by which metabolic patterns may be adapted quantitatively to the different needs of tissues or organelles while remaining qualitatively similar. This may result not only from the different catalytic properties of isoenzymes arising from the tissue-specific expression of particular loci or from the selective effect of activators or inhibitors on them, but also from the ability to vary independently the amounts of isoenzymes in different locations.

Correlations between the properties of isoenzymes predominant in certain tissues and the metabolic patterns of those tissues have been clearly demonstrated in several cases, although in other instances the putative relationship between properties and function has not been established. As in other areas of isoenzyme studies, the selective deletion of specific isoenzymes by mutation has been particularly useful in elucidating their metabolic roles.

Isoenzymes in carbohydrate metabolism

Although glycolysis is a universal metabolic pathway, the relative importance of the breakdown of glucose or glycogen to provide energy and of the reversal of these processes varies greatly from one tissue to another. In addition, connections between carbohydrate metabolism and the metabolism of amino acids and lipids are particularly important in the liver in contrast to other tissues. Almost all the glycolytic enzymes exist as isoenzymes and the occurrence of typically 'liver' or 'skeletal muscle' isoenzyme distributions is particularly noteworthy, several examples of which have already been mentioned.

In general, the properties of these liver- or muscle-specific variants are consistent with the trends of carbohydrate metabolism in the respective tissues. This is the case, for example, for the isoenzymes A and B of aldolase, predominant respectively in skeletal muscle and liver. Both isoenzymes will cleave fructose-1,6-diphosphate or fructose-1-phosphate, but the aldolase A of muscle shows a fifty-fold greater activity toward the diphosphate than towards fructose monophosphate. This is in keeping with its part in the glycolytic metabolism of skeletal muscle, since cleavage of fructose-1,6-diphosphate to triosephosphate is a key reaction in glycolysis. Aldolase B, the main isoenzyme of liver, shows no marked preference for the diphosphate substrate and this and other properties indicate that it is better adapted to utilization of fructose and to gluconeogenesis (Horecker, 1975). The substrate specificity of aldolase C is intermediate between those of the A and B isoenzymes, but the relationship between the properties of this isoenzyme and the metabolism of tissues in which it occurs, such as the brain, remains obscure, as it does in the case of other tissues and isoenzyme systems in which isoenzymes with intermediate properties predominate.

If glucokinase is considered to be a member of the hexokinase isoenzyme system, a striking correlation is evident between its properties and glucose utilization in liver, in which it is specifically localized. The higher K_m value of 2×10^{-2} mol l^{-1} for glucokinase, contrasting with the much lower value of 10^{-5} mol l^{-1} for hexokinase isoenzyme l in brain, is of the same order of magnitude as the glucose concentration of the blood. Since the hepatocyte is freely permeable to glucose, the rate of phosphorylation of glucose in liver cells is governed by the blood glucose level. In contrast, brain hexokinase can be assumed to be almost always saturated with substrate, ensuring efficient utilization of low blood glucose concentrations.

Although the group of low-K_m isoenzymes of hexokinase (excluding glucokinase) have tissue specific distributions, with brain and skeletal muscle showing distinctly different patterns, the metabolic roles of these isoenzymes have not been fully explained (Purich *et al.*, 1973). It has been suggested that the increase of hexokinase activity in skeletal muscle and some other tissues in response to administration of insulin is due to increased synthesis of isoenzyme II, whereas there is no similar response in tissues such as brain in which isoenzyme l is mainly present. This possibility illustrates a further potential means of selective regulation of metabolism by isoenzymes, and one which may also suggest a general function for those isoenzymes with catalytic properties which are not obviously adapted to specific metabolic needs; that is, the possibility of tissue-specific variation of the level of a particular catalytic activity.

The marked kinetic differences between the main pyruvate kinase isoenzyme of muscle and the liver isoenzyme may also be interpreted in terms of different metabolic functions. The predominant form of this enzyme in muscle exhibits a hyperbolic relationship between velocity and substrate concentration, whereas for the isoenzyme of liver the curve is sigmoidal. Fructose diphosphate is an activator of liver pyruvate kinase, but not of the muscle isoenzyme. Thus, the rate of reaction of liver pyruvate kinase is more responsive to changes in the concentration of its substrate, and also to the rate of production of fructose diphosphate at an earlier stage of glycolysis. A further difference between these two isoenzymes is that the pyruvate kinase of liver is inhibited by L-alanine, providing a means by which the rate of oxidation of carbohydrate can be reduced in favour of gluconeogenesis.

Other isoenzymes of carbohydrate metabolism with sigmoidal kinetics include those of phosphofructokinase. In this case, the muscle isoenzyme shows a more pronounced sigmoidicity than the liver form with fructose-6-phosphate as substrate. Phosphofructokinases are activated or inhibited by various metabolites, notably adenine nucleotides, and differences between the isoenzymes in these respects provide further opportunities for selective regulation of metabolism.

The significantly different catalytic properties of the H_4 and M_4 homotetramers of lactate dehydrogenase have been studied extensively in the search for inferences as to the physiological function of these isoenzymes. The H_4 isoenzyme is inhibited by excess pyruvate to a greater extent than the M_4 isoenzyme under certain conditions *in vitro*. On the basis of this observation, it has been suggested that rapid accumulation of lactate can occur in a tissue in which the main form of lactate dehydrogenase is M_4, whereas if H_4 is the predominant isoenzyme, its inhibition by excess pyruvate will prevent conversion of excess pyruvate to lactate. The ability of tissues to function anaerobically, i.e. in conditions leading to accumulation of lactate, is therefore presumed to be due to the presence in them of adequate levels of the M_4 tetramer (Cahn *et al.*, 1962).

Impressive support for the hypothesis that capacity for anaerobic metabolism and the presence of M_4 lactate dehydrogenase are correlated has been gathered from the study of the isoenzyme composition of muscles from avian species, in which a clear distinction can be drawn between muscles capable of sustained contraction and those which are active only intermittently. The former muscles which, like human cardiac muscle, operate aerobically contain the electrophoretically-faster lactate dehydrogenase isoenzymes, i.e. those composed wholly or mainly of H-subunits, while the latter, like human skeletal muscle, contain the slower isoenzymes (Wilson *et al.*, 1963). However, this generalization does not appear to be universally applicable since liver contains a large amount of the M_4 isoenzyme but has an aerobic pattern of metabolism. Furthermore, the differences between the LD_1 and LD_5 isoenzymes in their sensitivity to inhibition by excess substrate *in vitro*, on the basis of which their different physiological functions were postulated, are not so marked at concentrations of substrate and enzyme corresponding to those believed to prevail *in vivo* (Vesell and Pool, 1966). Therefore, an alternative proposal has been made that abortive ternary complexes formed between lactate dehydrogenase, pyruvate and the oxidized coenzyme, NAD^+, are responsible for enzyme inhibition within the cell, and that these complexes are more stable in the case of the H_4 isoenzyme than with M_4 (Kaplan *et al.*, 1968).

Complete agreement has not yet been reached as to how far the different patterns of aerobic and anaerobic metabolism of tissues reflect the catalytic characteristics of their predominant isoenzymes of lactate dehydrogenase; indeed, the metabolic advantages of these isoenzymes may lie in the opportunities they provide for the regulation of total lactate dehydrogenase activity through tissue-specific inhibitors of protomer aggregation (Schoenenberger *et al.*, 1980), rather than in their catalytic differences. Nevertheless, studies of lactate dehydrogenase represent important attempts to give functional significance to variations in isoenzyme composition between tissues.

Theories as to the functions of some other catalytically-distinct isoenzyme products of multiple gene loci are even less well developed. The MM and BB isoenzymes of creatine kinase differ to some extent in their quantitative catalytic properties, such as Michaelis constants for the various substrates of the forward or reverse reactions, and, as with lactate dehydrogenase, these catalytic differences between creatine kinase isoenzymes may provide indications of their physiological functions. For example, the MM isoenzyme may be better adapted to the rapid generation of ATP from creatine phosphate to provide energy for rapidly-contracting skeletal muscle. However, this must be regarded as a tentative hypothesis, although the association of MM isoenzyme with white skeletal muscle fibres, and perhaps more specifically with their myofibrils, lends some support to it.

Differential localization of isoenzymes within cells

The need for certain enzymic activities to be represented in both mitochondria and cytoplasm derives from the impermeability of the mitochondrial membrane to some metabolites, so that duplication of the reactions which generate these metabolites is necessary in order to make them available for both mitochondrial and cytoplasmic metabolism. The existence of distinct mitochondrial and extra-mitochondrial isoenzymes potentially can provide, in each of these subcellular compartments, an enzyme form with catalytic characteristics appropriate to the substrate concentrations and direction of reaction prevailing in that location. An example is provided by the part played by the isoenzymes of malate dehydrogenase in the metabolism of acetyl-CoA. This compound is generated within the mitochondrion, the inner membrane of which is impermeable to it, but is used in the synthesis of fatty acids by enzymes located in the cytoplasm. A net transfer of acetyl-CoA from mitochondrion to cytoplasm is achieved by the intra-mitochondrial conversion of acetyl-CoA and oxaloacetate to citrate by the enzyme citrate synthetase. Citrate diffuses into the cytoplasm where acetyl-CoA is regenerated by the action of ATP-citrate lyase. The oxaloacetate formed in this reaction is reduced to malate by the cytoplasmic isoenzyme of malate dehydrogenase. Malate is further converted to pyruvate which can re-enter the mitochondrion, where oxaloacetate is again formed by stages including the oxidizing action of mitochondrial malate dehydrogenase (Fig. 4.5).

This scheme suggests that consumption of malate is favoured in the mitochondrion and production of malate in the cytoplasm. The respective properties of the mitochondrial and cytoplasmic isoenzymes of malate dehydrogenase are consistent with this: the mitochondrial isoenzyme is inhibited by excess oxaloacetate but not by malate, i.e. it is better adapted for production of oxaloacetate from a relatively high concentration of malate, whereas the pattern of inhibition is reversed for the cytoplasmic isoenzyme, which is thus better adapted to catalyse the conversion of

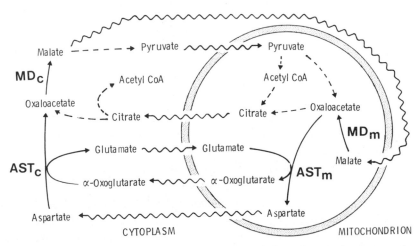

Fig. 4.5 *Patterns of cytoplasmic and mitochondrial metabolism, showing the effects produced by cytoplasmic (c) and mitochondrial (m) isoenzymes of malate dehydrogenase (MD) and aspartate aminotransferase (AST) working in opposite directions in the two subcellular compartments.*

oxaloacetate to malate. Since the inner mitochondrial membrane is also not freely permeable to oxaloacetate, the cytoplasmic and mitochondrial aspartate aminotransferase isoenzymes working in opposite directions effectively promote a transfer of oxaloacetate out of the mitochondrion as aspartate.

In other subcellular organelles the existence of isoenzymes with properties adapted to their microenvironments, such as the more acid pH optima of lysosomal isoenzymes compared with their extra-lysosomal analogues, can be regarded as further examples of the localized diversification of fundamentally similar processes which isoenzymic variation makes possible.

Functions of other multiple forms of enzymes

Multiple forms of enzymes which arise by post-translational modifications of a single gene product do not in general differ markedly in their catalytic characteristics. Therefore, the inferences as to the metabolic effects of such differences that have been so useful in suggesting possible physiological functions for true isoenzymes cannot usually be drawn in the case of these 'secondary isoenzymes'. Some multiple forms of enzymes, particularly those of membrane-bound enzymes, are artefactual, in the sense that their properties or even their existence depend on the methods used to extract them from the tissues. However, these properties of interaction with other cellular components and the nature of the substances involved may themselves provide evidence of specific adaptations of the enzyme molecules which are responsible for their localization and orientation in the membranes or other structures. Examples of structural modifications of basically similar enzyme molecules which result in or accompany their assumption of specific local roles in cellular metabolism that have already been mentioned include the microsomal and lysosomal β-glucuronidase variants, and the multiple supramolecular forms of cholinesterase.

Some post-translational modifications of enzyme molecules may produce changes in their catalytic characteristics, however, thereby introducing the possibility of different metabolic roles for the altered forms. Several forms of alkaline phosphatase in *E. coli* arise by post-translational modifications of a common primary structure. In the case of two of the forms, this modification consists of the removal of an amino-terminal arginine residue by proteolytic action within the bacterial cell. The transformation is accompanied by a decrease in the steady-state rate of hydrolysis of phosphate esters at low pH. Thus, the unmodified enzyme would appear to offer a physiological advantage for growth under more acidic, phosphate-deficient conditions (Schlesinger *et al.*, 1975). It seems possible that post-synthetic modifications of enzyme molecules producing functionally altered forms constitute a mechanism for the regulation and adaptation of enzyme activity in cells which is of quite general application.

Many of the differences between multiple forms of enzymes which do not arise at the level of the structural gene are due to variations in the composition of carbohydrate side-chains. The structures of the carbohydrate moieties of glycoproteins are determined by the action of specific enzymes, suggesting that differences in carbohydrate side-chains are of physiological significance. The appropriate structure of the carbohydrate moiety of some glycoproteins is probably important in ensuring their incorporation into various cellular structures, while experimental evidence has been obtained for the regulation of uptake of extracellular proteins by means of carbohydrate-specific receptors at cell surfaces. A general role has been suggested for the terminal sialic acid residues of some glycoproteins, in preventing recognition of circulating glycoproteins by the galactosyl-receptor sites of hepatocytes and thus prolonging the survival of the sialoproteins in the circulation (Morell *et al.*, 1971). Modification of the carbohydrate portions of multiple forms of enzymes may therefore offer additional possibilities for the selective regulation of enzyme activity by altering the effective life of enzyme molecules.

5 Multiple Forms of Enzymes in Phylogeny and Genetics

Since the patterns of reactions which constitute the fundamental processes of energy production, growth and reproduction are broadly similar across the whole spectrum of living matter, it is to be expected that individual enzymes catalysing these reactions in a given organism will have functional counterparts in other species, even those that are morphologically and physiologically remote. Where more specialized biochemical processes are concerned the distribution of analogous enzymes will be narrower, but here also similarities between enzymes exercising parallel functions in different species are to be expected. Although the various forms of enzymes with similar catalytic activities derived from different species are not isoenzymes by strict definition, comparisons of their structures and properties can provide valuable aids to taxonomy, as well as insights into the possible course of evolution of the enzymes themselves and of the species in which they occur. These opportunities were recognized early in the systematic study of iso-enzymes and have been actively exploited (Market and Møller, 1959; Watts, 1968; Masters and Holmes, 1974).

GENE DUPLICATION AND THE EVOLUTION OF MULTIPLE LOCI

Enzymes with comparable functions might have become distributed throughout living matter as a result of two different evolutionary processes. In the first, a particular catalytic activity may have arisen from two or more quite unrelated events, leading to the emergence of independent genes with products each capable of catalysing, for example, peptide-bond cleavage. Although convergent evolution of the separate genes may subsequently have increased the similarities between their

117

dependent enzymes, these would nevertheless be expected to remain distinct in structure, and perhaps in catalytic mechanism also, as is the case for the various classes of proteases.

The second evolutionary process which may be envisaged is the duplication of a single ancestral gene which determines the structure of a specific enzyme. The two genes thus produced can then be subjected to independent mutations, with corresponding modifications of their product enzymes. The existence of the differentially modified enzymes, or isoenzymes as they have now become, would favour the adaptation of patterns of metabolism to the particular needs of evolving organisms and the division of physiological function between specialized cells and organs. The operation of natural selection on the modified genes would ensure the distribution of favourable modifications throughout the species in which they arise and maintain them through subsequent evolution. Although the independent and potentially numerous mutations of the duplicated genes would inevitably result in divergence of the structures of their isoenzymic products and thus also of their properties, the residual similarities between them should be sufficient to indicate their common ancestry.

Gene duplication appears to be the most probable cause of the existence of multiple loci determining the structures of isoenzymes, not only within a single species but throughout the living world. The course of events leading to the emergence of particular loci can be traced by comparisons of some well-characterized and widely-distributed enzyme forms.

Isoenzymes of lactate dehydrogenase in various species

The isoenzymes of lactate dehydrogenase have marked advantages in phylogenetic studies: the enzyme itself is universally distributed; its isoenzymes from several species have been the subjects of purification and structural analysis; structural and genetic studies have confirmed the existence of multiple gene loci and the formation of hybrid isoenzymes, and post-translational modification does not complicate the interpretation of the distribution and properties of the isoenzymes to any significant extent.

The existence of two main genetic loci, A and B, determining the structures of distinct polypeptides of lactate dehydrogenase, M and H (or A and B), and the properties of the homo- and heterotetrameric isoenzymes assembled from these polypeptides, have already been described.

Homologous loci are active in fish, amphibians, reptiles and birds, as well as in man and other mammals. Reference has also been made to a third lactate dehydrogenase locus, C (or X) which gives rise to a characteristic tetrameric LD_X isoenzyme, but only in the primary spermatocytes of birds and mammals.

A further lactate dehydrogenase locus, E, is expressed in retinae of teleost fish from the beginning of retinal differentiation (Whitt, 1969). Since the B locus is simultaneously active in the same cells, hybrid BE isoenzymes are detectable (Fig. 5.1). Lactate dehydrogenase molecules with characteristics similar to those of isoenzymes determined by the E locus are also present in the livers of some families of teleost fish and probably indicate that the E locus is expressed in these organs also. In contrast to the A and B isoenzymes, for which similar electrophoretic mobilities are found between homologous isoenzymes from many fish, the E isoenzymes show marked species variation in this property (Shaklee *et al.*, 1973). Salmonid fish exhibit particularly complex patterns of lactate dehydrogenase isoenzymes as a result of duplication and modification of both the A and B loci, the unmodified (A and B) and modified (A′ and B′)

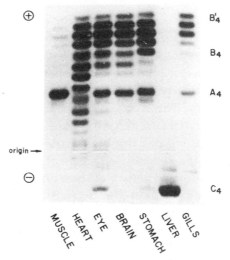

Fig. 5.1 *Isoenzymes of lactate dehydrogenase in extracts of tissues of the silver hake,* Merluccius bilinearis, *separated by starch-gel electrophoresis. The most anodal isoenzymes consist of homo- and heterotetramers of subunits determined by the A and B loci and by a variant gene at the B locus (B′). The most cathodal isoenzyme is the homotetramer composed of subunits determined by the locus designated C by the original authors and as E in the present text. Its product is particularly prominent in the liver of this species. Heteropolymers of the E (or C) subunits with other subunits occur anodally of the E_4 (C_4) band (From Shaklee* et al., *1973. By permission of Wistar Institute Press).*

loci all being expressed (Massaro and Markert, 1969). Since the E locus is also active, five isoenzymes consisting of A and A′ subunits, five with B and B′ subunits, and isoenzymes consisting of B, B′ and E subunits can be found.

The various homotetrameric isoenzymes of lactate dehydrogenase from a given species show considerable structural homologies, e.g. in amino acid composition and partial sequences. Peptide mapping of M(A), H(B) and X(C) subunits of mouse isoenzymes indicates greater structural similarities between the A and B subunits than between either of these and the C subunit (Chang *et al.*, 1979). However, the C polypeptide is rather more similar to the B than to the A subunit in amino acid composition and in structure-related properties such as stability to heat or concentrated urea solutions, although all three polypeptides are antigenically distinct. Resemblances in stability also exist between the E and B homotetramers of teleost fish, and this implied structural similarity is supported by immunological cross-reaction between the B and E isoenzymes (Shaklee *et al.*, 1973).

Evolution of lactate dehydrogenase genes

Enzymes with characteristics associated with the A isoenzyme of lactate dehydrogenase occur in a very wide variety of species, and their similarities of properties reflect structural homologies which become more apparent as enzymes from more closely related species are examined. The B, C and E isoenzymes similarly each have their homologues in a more or less wide range of species. These observations are consistent with the emergence of the various lactate dehydrogenase genes by successive duplications and modifications of a single ancestral gene. In spite of their subsequent divergence, the common ancestry of the present genes is demonstrated by the widespread ability of their product polypeptides to form hybrid molecules, both within and between species (Fig. 5.2).

Of the existing genes, the one determining the structure of the A (or M) polypeptide appears to be most closely representative of the ancestral gene; it is more active than the B locus in many vertebrate tissues, and the A locus is less variable than the B locus when the isoenzymes of various species of teleost fish are compared (Whitt, 1969). The duplication and mutation which gave rise to the B (or H) locus must have taken place at about the time of emergence of the vertebrates since of this phylum only the lamprey resembles invertebrates in possessing a single lactate dehydrogenase locus. The further lactate dehydrogenase genes appear to derive from modification of the B locus, to produce the E locus in teleost

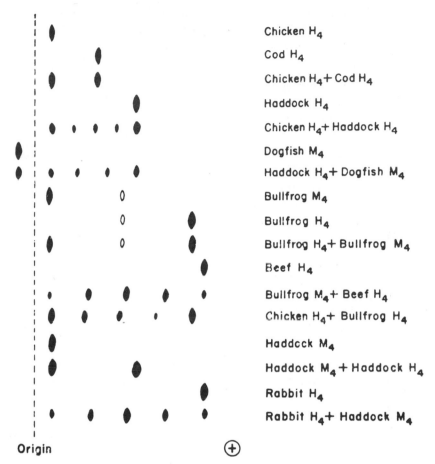

Chicken H₄

Cod H₄

Chicken H₄+ Cod H₄

Haddock H₄

Chicken H₄+ Haddock H₄

Dogfish M₄

Haddock H₄+ Dogfish M₄

Bullfrog M₄

Bullfrog H₄

Bullfrog H₄+ Bullfrog M₄

Beef H₄

Bullfrog M₄+ Beef H₄

Chicken H₄+ Bullfrog H₄

Haddock M₄

Haddock M₄+ Haddock H₄

Rabbit H₄

Rabbit H₄+ Haddock M₄

Origin

Fig. 5.2 *Diagram of the results of experimental hybridization of lactate dehydrogenase isoenzymes from a variety of species (From Salthe* et al., *1965. By permission of Macmillan, London.*

fish and the C locus in birds and mammals. The immunological cross-reactivity between the B and E polypeptides supports this postulated derivation, as does the close linkage between the B and C genes in pigeons (Zinkham *et al.*, 1969), contrasting with the location of the A and B genes on separate loci in man and other species. Duplication and subsequent differential modification at both A and B loci probably accounts for the existence of additional A′ and B′ genes in salmonid fishes.

Widely-distributed multiple forms of other enzymes

Some analogous enzymes appear to have been derived from distinct evolutionary processes, as has been mentioned in the case of various classes of proteases. A further example is provided by the two distinctive types of fructose diphosphate *aldolases*. Class I aldolases, tetrameric enzymes with lysine at the active site and carboxy-terminal tyrosine residues, occur in protozoa, plants and animals, while class II consists of the dimeric, metal-dependent aldolases of bacteria, yeast and fungi.

Aldolases of class II from various species show little variability (Lebherz and Rutter, 1969), but homo- and heterotetrameric forms of class I aldolases are widely distributed in vertebrate species, with some multiplicity also evident in invertebrates and plants (Lebherz and Rutter, 1969). As with lactate dehydrogenase, the existence of homologues of the mammalian A, B and C aldolase genes is taken as evidence of gene duplication and modification events associated with the differentiation of function between organs such as liver and muscle, but with conservation of the essential catalytic mechanism. As is also the case with lactate dehydrogenase, the existence of a fourth aldolase in salmonids is attributed to the tetraploidy of these fish.

The MM, MB and BB dimeric isoenzymes of *creatine kinase* found in mammalian tissues have their homologues in the tissues of other vertebrates such as birds and teleost fishes. Several additional forms of the enzyme have been detected in fish and birds, although the genetic origin of some of these forms is not yet clear. Two forms of CK-BB were found in 250 species of birds belonging to 15 different orders, except for a few Psittaciformes species (Scholl and Eppenberger, 1969). Although arginine kinase is responsible for phosphotransferase reactions involving ATP in invertebrates in general, creatine kinase is present in a number of related invertebrate species, and invertebrate creatine kinase cross-reacts immunologically with the enzyme from teleosts (Fitzsimmons and Doherty, 1970). Hybridization of the subunits of mammalian creatine kinase and echinoderm arginine kinase has been reported, indicating a high degree of homology of their structures and suggesting that these two enzymes derive from a common evolutionary precursor (Watts *et al.*, 1972).

As would be expected, the distinctive mitochondrial and cytoplasmic isoenzymes of enzymes such as *malate dehydrogenase* are widely distributed. The electrophoretic mobilities of the cytoplasmic isoenzymes from a large number of species of birds have been compared (Kitto and Wilson, 1966) and a similar comparison has been made of this isoenzyme from 32 marsupial species (Holmes *et al.*, 1974). The malate dehydrogenases from the majority of bird species, including all those from the order Galliformes, had identical electrophoretic mobilities but an enzyme form of distinctive mobility was present in birds belonging to the orders Charadiiformes (shore birds), and Apodiformes (swifts and humming birds) also possessed a unique form. These observations seem to confirm the unity of each of these orders, which has been open to question.

A similarly high degree of conservation of the structure of cytoplasmic malate dehydrogenase was found in the study of this isoenzyme in marsupials. However, the structure of mitochondrial malate dehydrogenase is less strongly conserved in both marsupials and birds, with variations in electrophoretic mobility existing between preparations of this isoenzyme from closely related species.

Three classes of human *alkaline phosphatase* can be distinguished on the basis of stability to heat, sensitivity to various inhibitors, and antigenicity. These are placental phosphatase, intestinal phosphatase, and phosphatases from tissues other than these such as liver, bone and kidney (Chapter 4). The properties of human alkaline phosphatases of the third group are reflected in the enzymes from homologous tissues of a variety of non-human mammals suggesting considerable conservation of the gene (or genes) responsible for this group of phosphatases. Similarly, intestinal alkaline phosphatases from several mammalian species have many characteristics in common, although inhibitor studies show a greater inter-species variation than is seen among non-intestinal phosphatases, indicating a correspondingly greater degree of variation in the structure of the inhibitor-binding site of the intestinal enzyme (Goldstein *et al.*, 1980).

The most marked differences in properties between human alkaline phosphatases are those distinguishing placental from non-placental phosphatases. However, in non-hominid species placental alkaline phosphatase is similar to liver phosphatase (Goldstein and Harris, 1979), while among Hominidae the resistance to heat inactivation and sensitivity to bromotetramisole, characteristic of human placental alkaline phosphatase, are only exhibited by placental phosphatases from chimpanzee and orangutan; furthermore, placental phosphatases from these two species are immunologically identical with the human placental enzyme (Doellgast and Benirschke, 1979). Therefore, the emergence of a specific placental alkaline phosphatase gene appears to have been a relatively late evolutionary event.

The problems of classifying the multiple forms of *esterases* of low specificity which are present within a single species as isoenzymes or as distinct but functionally related enzymes have already been mentioned. Furthermore, esterases are capable of undergoing a variety of post-translational modifications due to sialylation or aggregation which further complicate the interpretation of their heterogeneity. These problems are magnified when attempts are made to identify homologous forms and compare them between different species, and thus detract from the value for phylogenetic studies of what would otherwise be highly suitable enzyme systems for this purpose. Nevertheless, the wide distribution of esterases and the ease with which their activities can be demonstrated after zone electrophoresis have encouraged a large number of comparative studies. As well as electrophoretic mobility, molecular size (judged by migration through starch- or polyacrylamide-gels), relative activity with various substrates, and response to several inhibitors, have been used to classify the multiple enzyme zones present in many species (Masters and Holmes, 1974).

At least five groups of aliesterases, which preferentially (but not exclusively) act on straight-chain aliphatic esters and are distinguishable from cholinesterases by their resistance to inhibition by eserine, have been identified in vertebrates by these criteria. Although most of the groups are to some extent heterogeneous, genetic evidence suggests that this is probably due in each case to post-genetic modification of a single gene. The differential distribution of apparently homologous enzymes amongst vertebrate species suggests a history of evolution and in-dependent mutation of multiple gene loci. Similarly, three groups of arylesterases, with a preferential action on esters of aromatic carboxylic acids, have been described. Two apparently homologous groups occur widely in both vertebrates and invertebrates, whereas the third group is of more limited distribution.

Homologues of the *carbonic anhydrase* isoenzymes I and II from human red blood cells have been found in all species of placental mammals so far examined. Comparisons of the amino acid sequence of isoenzymes from different species show more extensive identity between homologous isoenzymes from different species than between the two isoenzymes from a single species. The percentage of identical residues is greater than 80 when the primary structures of human, ox and rabbit carbonic anhydrase isoenzymes I are compared, with 94% of identical residues in the case of human and rhesus monkey isoenzymes. Similar degrees of identity exist between the isoenzymes II of these four species (Tashian, 1977). In contrast, isoenzymes I and II from any one of these species have approximately 60% of their sequences in common.

Although few studies have so far been made, only a single isoenzyme (corresponding to isoenzyme II of higher mammals) was found in the erythrocytes of the red kangaroo, suggesting that the gene duplication giving rise to isoenzymes I and II took place at an early stage in the evolution of placental mammals (Tashian, 1977), but at a relatively late stage in vertebrate evolution. The close genetic linkage of the loci determining mammalian carbonic anhydrases I and II also suggests a recent common origin.

The carbonic anhydrase of avian erythrocytes resembles the mammalian isoenzyme II in its high specific activity and also in its pronounced esterase activity with substrates such as p-nitrophenyl acetate. Evidence for the evolution of the carbonic anhydrase isoenzymes of vertebrate erythrocytes from the avian enzyme, or from a common precursor similar to it, has been sought in the comparison of amino acid sequences at the carboxyl termini (Tashian, 1977). Although preliminary, these data suggest that one or more point mutations might have changed the C-terminal serine of the avian enzyme into the lysine, arginine or valine residues of mammalian isoenzymes I, while the terminal phenylalanine of isoenzyme II might be the result of deletion of the adjacent serine from the end of the avian enzyme.

A third isoenzyme of carbonic anhydrase with lower hydrase activity than the other isoenzymes and with low or absent esterase activity has been identified in skeletal muscle of some mammals; an enzyme found in chicken muscle may be homologous with the mammalian isoenzyme III. If structural studies support this conjecture, the evolutionary pathway leading from an ancestral enzyme to the present forms of carbonic anhydrase may be presumed to have branched at an early stage, one branch leading by successive duplications to the avian erythrocyte enzyme and the isoenzymes I and II of mammals, the other to the carbonic anhydrase III of mammalian and avian muscle (Tashian, 1977).

GENETICS OF ISOENZYMES

The multiple gene loci which have been conferred on a particular species by the processes of mutation and natural selection are expressed in each of its members as a characteristic complement of isoenzymes. On this common pattern are superimposed individual modifications due to allelic variation at one or more of the loci, which are inherited in accordance with Mendelian laws. In contrast to the inherited morphological traits of classical genetics, however, allelozymes are inherited codominantly; i.e., in an individual heterozygous at a given locus, both alleles are expressed as their characteristic allelozymes. Thus, the individual isoenzymic phenotype closely mirrors its genotype, and the isoenzyme pattern of the heterozygote is the sum of the patterns seen in the respective homozygotes, with the addition of any heteropolymeric isoenzymes which may arise from the association of genetically distinct polypeptide subunits (Fig. 2.3).

Although newly-developed techniques of direct analysis of DNA sequences, restriction endonuclease analysis and cloning have to some extent reduced the importance of isoenzymes as indirect indicators of genetic change, the expression of genes and the functional characteristics of their products remain essential concerns of the science of genetics. In this respect studies of inherited enzyme variants and the genetic changes from which they derive are complementary.

The widespread and systematic application of the techniques of isoenzyme analysis to many invertebrate and vertebrate populations, including human populations, has revealed that the frequency and extent of allelic variation at many enzyme-determining loci is considerable (Lewontin, 1974; Harris, 1975). The zymogram technique, in which specific staining for enzyme activity follows separation by zone electrophoresis, is particularly useful in these studies because of its sensitivity and its applicability to impure extracts.

Further characterization of the allelic variants of a single enzyme often fails to demonstrate significant differences in catalytic properties between them, while individuals possessing different allelozymes have apparently identical metabolic processes.

The existence of this large body of stable but apparently purposeless genetic variation has prompted opposing 'selectionist' and 'neutralist' interpretations. In the selectionist view, almost all variation is maintained by the action of natural selection, although the advantages of one genotype over another in a particular environment may be so subtle as to escape detection. Furthermore, since the elimination of relatively unfavourable mutations might take many generations, both preferred and less favoured genes can be expected to be present in the population for long periods. On the other hand, neutralists maintain that the differences between many gene products are so slight that their parent genes escape the pressures of natural selection, changes in the frequencies of such neutral alleles over successive generations being merely the result of random genetic drift.

Evidence for or against these two theories has been sought in comparisons of the observed frequency of heterozygosity with the predictions of the neutralist theory or the expectations of the selectionist view. For example, invertebrates are more at the mercy of changes in their environments than are vertebrates, with their more advanced regulatory mechanisms. Invertebrates, therefore, might be expected, from a selectionist viewpoint, to have a greater range of genetic options with which to respond to environmental challenges than vertebrates would require. Although this type of generalization is broadly supported by observation, calculated frequencies of heterozygosity have also provided results consistent with neutralist expectations.

Attempts have also been made to modify the frequency of mutant alleles in a closed population and then to observe whether natural selection acts to restore the original frequency. Introduction of fruit flies (*Drosophila buzzattii*) carrying mutant genes at the three loci determining esterase-2, pyranosidase and alcohol dehydrogenase-1 into an isolated population of the flies over a short period approximately doubled the frequencies of the mutant genes. However, after introduction of the mutant genes ceased, the frequency of each gene in the population reverted to its original value, though at different rates (Barker and East, 1980). These results are consistent with natural selection acting differentially on the mutant genes.

Although no conclusive choice can be made between the neutralist and selectionist views of the existence of multiple allelozymes – indeed, it might be questioned whether a single explanation need be applicable to all examples of this phenomena – there can be no doubt of the disadvantages incurred by the inheritance of the rare alleles which account for many human diseases. Even when the product of the mutant allele is a demonstrable or presumed an inactive 'allelozyme', these conditions can still be regarded as the extreme end of a spectrum of isoenzymic variation.

Isoenzymes and genetic disease

The functional effects of the inheritance of variant enzymes in man can vary from the undetectable, through effects that are only apparent when their possessor is challenged by some event such as the administration of a specific food or drug, to crippling or even fatal disease. These graduations are related to the place which the affected enzyme occupies in metabolism and the nature and functional consequences of the alteration in structure of the enzyme molecule.

The function of *cholinesterase* in serum is unknown; virtually complete absence of the active enzyme, such as occurs in individuals who are believed to be homozygous for a rare 'silent' allele, is compatible with normal health. However, administration of the muscle relaxant suxameth-onium (succinyl dicholine; scoline) results in a prolonged period of apnoea instead of the usual brief effect. Varying degrees of sensitivity to this drug are shown by patients who are homozygous for alleles determining various uncommon isoenzymes of serum cholinesterase, or in whom different uncommon alleles are paired.

Apart from the silent allele, which produces no active enzyme, a rare allele has been identified which determines an isoenzyme with an increased resistance to the competitive inhibitor dibucaine compared with the usual isoenzyme (the 'atypical' isoenzyme), while another allele determines the 'fluoride resistant' isoenzyme with increased resistance to inhibition by this ion (Kalow and Genest, 1957; Harris and Whittaker, 1961) (Fig. 3.2). Both these uncommon isoenzymes have greater Michaelis constants than the usual form. Therefore, it appears that the alterations in molecular structure which distinguished them from the usual isoenzyme involve the active centre, reducing its affinity for substrates or inhibitors and thus the catalytic efficiency of the enzyme, since the sera of patients with the atypical or fluoride-resistant isoenzymes exhibit, on average, less cholinesterase activity than sera containing the usual isoenzyme. Direct evidence of a change in amino acid sequence near the active centre has been obtained by peptide mapping after partial proteolysis of the usual and atypical cholinesterases labelled with radioactive di-isopropyl fluorophosphonate (Muensch *et al.*, 1978). Expression of the various possible homozygous and heterozygous combinations of normal and uncommon alleles throughout the population gives rise to an almost continuous gradation of serum cholinesterase activity, from zero to normal levels, and to corresponding degrees of sensitivity to suxameth-onium. Inheritance of greater than normal activity of serum cholines-terase has also been described: it is apparently due to increased amounts of the usual isoenzyme, with a lower than normal sensitivity to suxameth-onium (Delbrück and Henkel, 1979).

Almost the full range of effects of allelic variation of an enzyme is shown by erythrocyte *glucose-6-phosphate dehydrogenase*, i.e. there are apparently no adverse consequences of the possession of some unusual variants whereas others cause moderate or severe haemolytic anaemias, either in response to some stress such as drugs or infection, or chronically present.

Deficiency of glucose-6-phosphate dehydrogenase is the most common disease-producing enzyme deficiency of human beings (Beutler, 1978). The absence of a directly fatal disease associated with deficiency of the enzyme can be attributed to its ancillary, rather than essential, role in the metabolism of red blood cells. Although the major part of the energy of red cells is generated by the Embden–Meyerhof pathway of glucose metabolism, a significant fraction (normally about 10%) of glucose is metabolized by the hexose-monophosphate shunt pathway, the first enzyme of which is glucose-6-phosphate dehydrogenase. Apart from the pentoses used in nucleotide synthesis, this pathway generates NADPH which maintains glutathione in its reduced state. In turn, reduced glutathione preserves the active sulphydryl groups of red-cell enzymes in their reduced form and protects haemoglobin against oxidation. The severity of the haemolytic anaemias associated with allelozymes of glucose-6-phosphate dehydrogenase is related to their ability to fulfill the normal functions of the enzyme.

Allelozymes of glucose-6-phosphate dehydrogenase have been detected and characterized by the typical methods of isoenzyme analysis: decreased catalytic activity, altered Michaelis constants for the substrates glucose-6-phosphate and NADP and reactivity with substrate analogues, changed electrophoretic mobility and other charge-dependent properties, and altered stability (Beutler, 1978).

As well as the production of a less active or less catalytically-efficient enzyme, a functional deficiency can result from a reduced stability of the enzyme in the red cells. Since enzymes and other proteins are not synthesized in these cells, deficiencies cannot be made good by increased enzyme synthesis during the life of the cells, and the consequent decline in glucose-6-phosphate dehydrogenase activity is accelerated in the cells of subjects with some allelozymes. The alteration in the functional characteristics of the allelozymes is correlated with the degree of severity of associated haemolytic anaemias, and with sensitivity to drugs such as antimalarials which may precipitate haemolytic crises. As would be expected, there is no correlation between changes in electrophoretic mobility and haemolytic disease.

Glucose-6-phosphate dehydrogenase is a polymeric enzyme, possibly dimeric *in vivo*, and hybrid isoenzymes have been prepared *in vitro* between human allelozymes and between rat and human enzymes. The structural difference between the most common allelozyme, glucose-6-phosphate dehydrogenase B, and the electrophoretically distinct A isoenzyme which is present at normal levels of activity in 20% of American negroes, has been shown to consist of a single amino acid substitution, with an aspartic acid residue in the A form replacing an asparagine of the B isoenzyme. A single amino acid substitution has also been shown to account for the different structure of the Hektoen variant, and it seems probable that almost all the glucose-6-phosphate dehydrogenase allelozymes will prove to derive from point mutations, each involving a change of only a single base in the structural gene.

When a particular type of enzyme activity is due to the actions of isoenzymes produced by multiple gene loci, the metabolic effects of mutation at only one locus are confined to those tissues or organs in which the affected locus is expressed. This is the case in the glycogen storage diseases types V and VI. The cause of both diseases is a deficiency of *phosphorylase* (the enzyme which catalyses the breakdown of the α-1,4 linkages of the glucose chains of glycogen), with the result that glycogen accumulates in the tissues. However, in type V (McArdle's) disease, muscle phosphorylase alone is affected and glycogen accumulation occurs only in this tissue. Inability to mobilize glycogen leads to muscle cramps on exercise. A mutation at the locus determining liver phosphorylase seems to be responsible for type VI (Hers') disease, in which the liver is enlarged by accumulation of glycogen and moderate hypoglycaemia occurs.

The inheritance of a catalytically ineffective form of a widely-distributed enzyme would be expected to have equally generalized metabolic consequences and this is indeed the case when the affected enzyme is a component of the lysosomes, since these organelles are present in all cells with the possible exception of erythrocytes. The degradation of many types of macromolecules takes place in lysosomes; therefore, deficiency of a lysosomal enzyme results in accumulation of its specific substrate and more than 30 such *lysosomal storage diseases* are now recognized (Stanbury *et al.*, 1978). Some of these enzyme defects have already been described as evidence for the separate identities of certain lysosomal and non-lysosomal isoenzymes (Chapter 4). Although the metabolic consequences of lysosomal storage diseases are general, their severity does vary from one tissue to another depending on the rate of production of the non-degradable substrate: in the gangliosidoses, for example, the main clinical abnormality is progressive degeneration of the central nervous system, due to the accumulation of gangliosides which are important constituents of the cells of the grey matter. Non-neural tissues contain relatively little of these substances, but accumulation of acid mucopolysaccharide in the reticuloendothelial system in liver, spleen and bone marrow also takes place in generalized gangliosidosis.

The anatomical abnormality of cells in lysosomal storage diseases, with their grossly enlarged lysosomes, presumably mirrors their disturbed function. Although rupture of lysosomes and the release of their potentially destructive hydrolytic enzymes has not been observed in these diseases, the distortion of the cellular architecture by the enlarged lysosomes appears to be sufficient to account for the accompanying functional disturbances, especially in the central nervous system where structure and function are so closely related.

Virtually no active enzyme is produced in the fatal or severely disabling inborn errors of metabolism such as the lysosomal storage diseases; therefore, characterization of the abnormal allelozyme is not possible, although the presence of immunologically recognizable inactive 'allelozymes' (cross-reacting materials) has been demonstrated in several enzyme deficiencies. Equally, it is not possible to confirm that the same mutation is responsible for all cases of a particular disease, and evidence that different mutations can indeed give rise to similar clinical manifestations has been obtained in a number of cases. Two mucopolysaccharides, MPS IIIA and MPS IIIB, also known as Sanfilippo syndromes A and B, are both characterized by the accumulation of the mucopolysaccharide, heparan sulphate, and are clinically indistinguishable. However, the deficient enzyme in Sanfilippo A is heparan N-sulphatase, whereas the B syndrome is due to a deficiency of N-acetyl-α-D-glucosaminidase (McKusick *et al.*, 1978). Thus, these diseases are the result of separate mutations at loci which determine two entirely distinct enzymes.

The converse situation, in which different mutations at a single locus and the production of distinct allelozymes are associated with a variety of signs and symptoms, has already been noted with respect to the less serious abnormalities arising from allelism at the loci determining serum cholinesterase and erythrocyte glucose-6-phosphate dehydrogenase.

Among the more serious deficiency diseases, two further conditions included in the group of mucopolysaccharidoses, MPS IH (Hurler's syndrome) and MPS IS (Scheie's syndrome), have quite dissimilar clinical features; the marked mental deterioration, enlarged spleen and liver and bony deformities of Hurler's syndrome contrasting with the normal development and minor deformities of the extremities associated with Scheie's syndrome. In both conditions, however, heparan and dermatan sulphates accumulate because of a deficiency of the lysosomal enzyme, α-L-iduronidase.

The most probable explanation of the differences between the Hurler and Scheie syndromes is that they represent different allelic modifications at a single iduronidase locus, and that patients with one or other of these diseases are homozygous with respect to the corresponding allele

(McKusick *et al.*, 1978). This interpretation is supported by the occurrence of a disease of intermediate severity, in accordance with the expectation that the Hurler and Scheie alleles might come together in a heterozygous individual. The different severities of the two syndromes are presumably related to the degree to which the function of the allelozymes is impaired, but the low iduronidase activity which therefore might be expected to be present in Scheie cells has not been demonstrated. Failure to discover an enzyme with altered properties in this and other examples of less serious storage diseases may be attributable to the analytical methods employed in studies *in vitro*. Short-chain substrates such as glycosides of *p*-nitrophenol or 4-methyl umbelliferone are often used because of the sensitivity with which the aglycone can be detected after hydrolysis. However, mutation may so alter the substrate specificity of an allelozyme that its activity, or absence of activity, towards these substrates does not reflect its function *in vivo*.

A correlation between the level of activity of the affected enzyme, glucocerebrosidase (β-D-glucosidase), and the severity of disease has been found in Gaucher's disease, in which glucocerebroside accumulates in spleen and other tissues. Patients with less severe type 1 diseases have been shown to have only 12–14% of the normal activity in their tissues, while those with the most serious type 2 disease had virtually no activity (Brady, 1978). These observations are consistent with the view that different mutations at a single locus are the causes of these two types of disease. The residual enzyme present in type 1 disease appears to differ from the normal glucocerebrosidase in affinity for its substrate and in heat stability (Brady, 1978). A third variant of Gaucher's disease of intermediate severity, type 3, may be due to a third mutation or may represent heterozygosity for two abnormal alleles.

A single mutation may affect more than one isoenzyme, by altering the structure of a single polypeptide subunit which is a component of a homopolymeric isoenzyme and also of one or more heteropolymeric isoenzymes. The metabolic consequences of such a mutation may be different from those resulting from the modification of a subunit present in only one type of enzyme molecule. An example of similar, but not identical, diseases arising in this way is provided by two GM_2-gangliosidoses, Tay-Sachs (Type I) and Sandhoff's (Type II) diseases. In both these lysosomal storage diseases accumulation of the sphingolipid GM_2 in brain (Type I) and also in viscera (Type II) results in degeneration of the nervous system, blindness and death. The deficient enzyme in each case is *N*-acetyl-β-D-hexosaminidase. Two forms of the enzyme, hexosaminidases A and B, are normally present in all tissues other than red blood cells. Hexosaminidase A has a greater net negative charge than B and therefore migrates further towards the anode on electrophoresis. It is also more readily denatured by heat than hexosaminidase B. An additional more negatively charged minor component, hexosaminidase S, is also detectable in normal tissues. The activity of hexosaminidase A is grossly reduced in Tay-Sachs disease whereas that of hexosaminidase B is normal or even increased in brain tissue. In Sandhoff's disease, on the other hand, both hexosaminidase A and B are deficient, although hexosaminidase S is more prominent (Fig. 5.3).

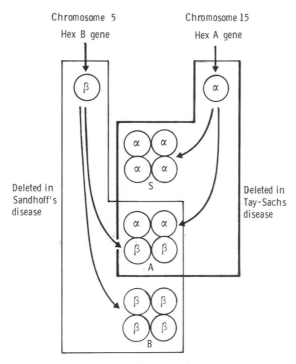

Fig. 5.3   *Diagrammatic representation of the occurrence and effects on hexosaminidase isoenzymes of the genetic defects in Sandhoff's and Tay-Sachs diseases.*

The postulated tetrameric structure of hexosaminidase has already been described (Chapter 4). The three isoenzymes represent homo- and heteropolymers of two dissimilar subunits, α and β, such that hexosaminidases B and S are the homotetramers β_4 and α_4, respectively, and hexosaminidase A is the heterotetramer, $\alpha_2\beta_2$ (Beutler and Kuhl, 1975).

The isoenzyme changes in the two lipidoses are accounted for if the mutation responsible for Tay-Sachs disease affects the α-subunit, so that hexosaminidase A cannot be formed; in this case, the excess β-subunits are available to form hexosaminidase B, which may be present in increased amounts. Similarly, if in Sandhoff's disease the formation of β-subunits is impaired, both hexosaminidases A and B are deficient but an excess of the α-subunit can give rise to an increased level of hexosaminidase S (Fig. 5.3). Isoenzyme A is more active on GM_2-sphingolipid than the homopolymers, which cannot compensate for its loss.

Before leaving the lysosomal storage diseases, the mucolipidoses types II and III may be mentioned. The biochemical features of these disorders include high activities of various lysosomal hydrolases in blood or urine, or in the medium bathing cultured cells. The activities of the corresponding enzymes in lysosomes are low (McKusick et al., 1978). A possible explanation of these findings is that the process by which these enzymes are sequestered in the lysosomes is defective, i.e. that the diseases are due to faulty post-translational enzyme modification. However, other explanations such as leakage of enzymes from lysosomes cannot be excluded.

Diagnosis of hereditary diseases and identification of carriers

When an abnormality of metabolism is due to the presence of an allelozyme with properties different from those of the usual isoenzyme, characterization of the variant enzyme may help to predict the probable extent of its metabolic consequences in the affected individual, and in others in whom a similar variant may be recognized. Examples of such clinical applications of isoenzyme studies include the assessment of risk of complications of anaesthesia due to inherited variants of serum cholinesterase, and, more importantly, the characterization of disease-associated allelozymes of erythrocyte enzymes such as glucose-6-phosphate dehydrogenase and phosphohexose isomerase. Since, typically, no active allelozyme is produced in the more serious inborn errors of metabolism, isoenzyme characterization has no part in arriving at a diagnosis. Diagnosis is established definitively by the absence of the specific enzymic activity, or is based on demonstration of increased levels of one or more metabolites derived from the affected metabolic pathway.

No effective treatment is available for the majority of inherited diseases, so that the purpose of diagnosis is to forecast the probable course of the disease and to eliminate other possible causes of the signs and symptoms. Exclusion from the diet of the substance which cannot be metabolized is effective in the treatment of phenylketonuria and galactosaemia, and early and accurate diagnosis is needed in these cases so that this type of therapy can be instituted promptly. The possibility of treating lysosomal storage diseases by enzyme replacement is being actively explored. In general, however, the only practicable approaches to the more serious diseases are preventative, in identifying carriers who can be warned of the risks of conceiving an affected child, and in attempting to diagnose the disease *in utero* so that abortion of an affected foetus can be considered.

Prenatal diagnosis is now possible in about forty genetic abnormalities, including several lipidoses and glycogen storage diseases. Diagnosis is based on an absence of the critical enzyme activity from fibroblasts obtained by amniocentesis and cultured *in vitro* to obtain an adequate mass of cells for analysis. As already mentioned, the absence of an active allelozyme usually precludes the use of isoenzyme characterization in diagnosis, but specific assay of hexosaminidase A by heat inactivation or electrophoresis in extracts of cultured amniotic cells distinguishes the selective loss of the A isoenzyme in Tay-Sachs disease from what may otherwise appear to be a slightly reduced or even normal total activity (O'Brien *et al.*, 1971).

Many hereditary diseases are inherited in a recessive manner and their effects are only expressed fully in individuals who are homozygous for the disease-determining mutant allele, or in whom two different mutant alleles of similar effects have come together at the same locus. An example of the latter possibility has been found in a family in which each parent was heterozygous for a different variant allele at the locus determining phosphohexose isomerase. Pairing of the variant alleles in the offspring resulted in haemolytic disease which proved fatal in one child, but which was treated successfully after prenatal diagnosis in a second (Whitelaw *et al.*, 1979). The amount of enzyme which is produced by the normal allele is usually sufficient to protect heterozygous individuals from the worst effects of enzyme deficiency. However, the normal isoenzyme may be quantitatively insufficient to prevent the appearance of some manifestations of metabolic abnormality in conditions of stress, e.g. when an abnormal load of the compound whose metabolism is affected by the mutation is administered. Where the mutant allele produces little or no active enzyme, the average enzyme activity in heterozygotes is lower than that in homozygotes for the normal allele, although the degree of overlap between the two populations in this respect may be too great to allow a particular individual to be assigned with certainty to one or other group solely on the basis of measurements of total enzyme activity. If the mutant allele itself produces an active but altered isoenzyme, detection of the abnormal isoenzyme provides evidence of the heterozygous state, but this rarely occurs in carriers of serious diseases.

Chromosomal assignments of enzyme-determining genes

Chromosomal mapping by the methods of classical genetics, in which the linkage of genes in controlled genetic crosses is determined, is of limited applicability to man, because of the rarity of some mutants, long generation times and few offspring of each mating. However, the technique of somatic cell hybridization (p. 29) has proved to be particularly valuable in mapping human chromosomes, and more than 75 gene loci determining human enzymes have now been assigned to chromosomes in this way (McKusick and Ruddle, 1977; Shows, 1977). The retention or loss of particular enzymes by different somatic-cell hybrids and the association of these events with the appearance or disappearance of other chromosomal markers enable the identity of the chromosomes which carry the enzyme-determining genes to be deduced. Human chromosomes, rather than those of rodent origin, tend to be lost during fusion of human and rodent cells, so that the technique can be used to map the human chromosomes. Some limitation on the scope of the method is imposed by the tendency of enzyme activities associated with the differentiation of cells to be lost during culture, although enzymes catalysing the more fundamental metabolic processes continue to be expressed.

Some tentative generalizations can be made about the location of multiple loci determining particular sets of isoenzymes on human chromosomes (Fig. 2.2). Separation on different chromosomes seems to be general. In ten instances the loci determining individual isoenzymes are on different chromosomes: acid phosphatase loci 1 and 2 are carried on chromosomes 2 and 11, respectively; enolase loci 1 and 2 on chromosomes 1 and 12; hexosaminidases A and B loci on chromosomes 15 and 5; the soluble and mitochondrial isoenzymes of isocitrate dehydrogenase are determined by loci carried on chromosomes 2 and 15; the lactate dehydrogenase A (or M) locus is on chromosome 11 and the B (or H) locus on chromosome 12; the cytoplasmic and mitochondrial malate de-hydrogenase loci are respectively on chromosomes 2 and 7; the loci of peptidases A, B, C and D are on chromosomes 18, 12, 1 and 19; the three phosphoglucomutase loci are located on chromosomes 1, 4 and 6, those for the two isoenzymes of superoxide dismutase on chromosomes 21 and 6, and mitochondrial and cytoplasmic thymidine kinase loci on chromo-somes 16 and 17 (McKusick and Ruddle, 1977). Rather few examples are known in which multiple loci determining the members of a family of isoenzymes are located on the same chromosome: the loci for human salivary and pancreatic amylases are both on chromosome 1, as are the loci which respectively determine soluble and mitochondrial guanylate kinases. Two adenylate kinase isoenzyme loci are on chromosome 9, with a third locus on chromosome 1 (McKusick and Ruddle, 1977).

The separation of multiple loci on different chromosomes may be a reflection of a longer evolutionary history. Some loci which are separated in man are linked in lower animals but others, such as the two amylase isoenzyme loci, are closely linked in species as evolutionarily distant as man and the Chinese hamster (Shows, 1977).

Particular interest attaches to those enzyme loci which, in man, are carried on the X chromosome. When these loci are the targets of damaging mutations, the affected alleles and the diseases which they cause are inherited in an 'X-linked recessive' manner. Since males have only one X chromosome, inheritance of an abnormal allele cannot be compensated for by a paired normal allele; therefore, sufferers from the disease are almost always males. Females are only affected in the rare eventuality that both X chromosomes carry the abnormal allele. Glucose-6-phosphate dehydrogenase and α-galactosidase (a lysosomal enzyme) are among the potentially disease-causing enzyme loci assigned to the human X chromosome.

The observation that females do not have a greater activity of X-linked enzymes than males in spite of the two X chromosomes that they possess has led to the formulation of the Lyon hypothesis (Lyon, 1961). This postulates that one of the pair of X chromosomes in each female cell is inactivated early in embryogenesis. The inactivation of either the maternal or the paternal X chromosome could take place at random; descendants of a cell in which the paternal X chromosome remained active would have only paternal X chromosomes, and vice versa. Thus, the individual would become a genetic mosaic for X-linked characteristics. Support for the Lyon hypothesis has been obtained from studies on the allelozymes of the X-linked enzyme, glucose-6-phosphate dehydrogenase. The most common variant of this enzyme is the B isoenzyme but the electrophoretically faster A form is also relatively common in black populations so that members of such populations are frequently heterozygous with respect to the alleles determining these two isoenzymes. Clones of cells derived from individual skin cells of heterozygous individuals were found to contain either the A or B isoenzymes, but not both, in accordance with the predictions of the Lyon hypothesis (Davidson *et al.*, 1963).

6 Multiple Forms of Enzymes in Ontogeny

Multiple gene loci and their dependent isoenzymes provide means for the adaptation of metabolic patterns to the changing needs of different organs and tissues in the course of normal development, or in response to environmental change. Pathological changes, also, may be associated with alterations in the activities of specific isoenzymes. Both physiological and pathological alterations in isoenzyme patterns are important, not only for the insights they provide into normal and abnormal metabolism, but also in the use of isoenzyme studies in medicine. Changes in the characteristics and distribution of multiple forms of enzymes which are not under genetic control occur in some cells as an accompaniment of ageing.

CHANGES DURING NORMAL DEVELOPMENT

Data on the electrophoretic patterns of 50 systems of multiple forms of human enzymes, including 29 in which multiple gene loci are involved, have been collected by Edwards and Hopkinson (1977), and similar studies have been made in other animal species and in plants (Scandalios, 1974). Although the nature and time-course of the changes in homologous enzymes often differ between species, the distribution of forms at a particular moment in ontogeny may be assumed to reflect the needs of the developing tissues, especially in the case of those isoenzymes determined by multiple gene loci (Fig. 6.1). Enzymes determined by single loci are generally present in both foetal and adult tissues.

Isoenzymes determined by multiple gene loci

Lactate dehydrogenase isoenzymes

As in so many other aspects of isoenzyme studies, the lactate de-

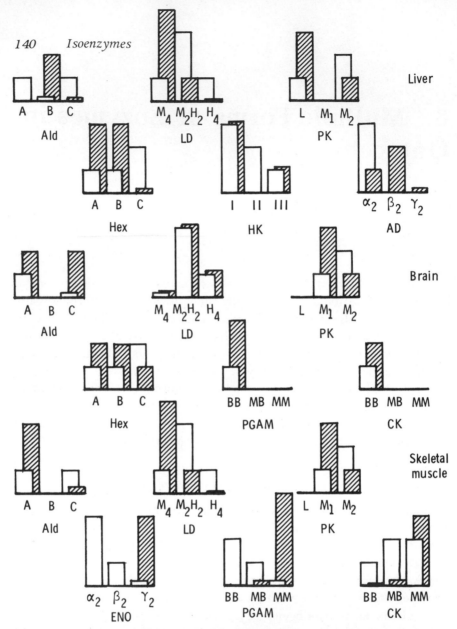

Fig. 6.1   *Differences in the patterns of distribution of nine isoenzyme systems between human foetal and adult tissues. The patterns of foetal liver, brain and skeletal muscle are shown as open bars, with hatched bars for the corresponding adult tissues. In these examples the adult and foetal patterns are generally qualitatively, as well as quantitatively, different. In many other instances the relative proportions of the isoenzymes are not altered so markedly during development although quantitative changes occur. AD, alcohol dehydrogenase; Ald, aldolase; CK, creatine kinase; ENO, enolase; Hex, hexosaminidase; HK, hexokinase; LD, lactate dehydrogenase; PGAM, phosphoglyceromutase; PK, pyruvate kinase. (Based on data of Edwards and Hopkinson, 1977).*

hydrogenase isoenzyme system has been the focus of much of the interest in isoenzyme changes in ontogeny, beginning with the pioneering observations of Markert and Møller (1959).

Knowledge of changes in lactate dehydrogenase isoenzyme patterns during early embryogenesis, i.e. between the stages of fertilization and implantation of the embryo, has been derived mainly from studies in the mouse (Brinster, 1979). The activity of lactate dehydrogenase in mouse embryos is initially very high but falls steeply by the time of implantation, which occurs on the fifth day after fertilization. The high activity in the earliest stages is almost exclusively due to the electrophoretically most-anodal LD_1 isoenzyme, implying that only H (or B)-type subunits are synthesized. The isoenzyme pattern changes completely after implantation and M (or A) subunits are produced almost exclusively, so that LD_5 becomes the predominant isoenzyme. Although there are quantitative differences in other species, B subunits appear to be most abundant in oocytes and in the early cleavage stages of the embryo, except possibly in the rabbit.

Since these changes in isoenzyme type may be presumed to have a functional basis, a correlation has been sought between the metabolic requirements of the embryo and its constituent isoenzymes. Although the relationship between isoenzyme type and metabolic pattern is still controversial in this as in other tissues, the observations in mouse embryo appear to be in accordance with the generalization that LD_1 predominates in tissues with an essentially aerobic metabolism such as exists in the embryo between fertilization and implantation.

Changes in the distribution of lactate dehydrogenase activity among the various isoenzymes during later stages of development have also been demonstrated in numerous studies. The electrophoretically-slower iso-enzymes predominate in various tissues in this species, with the more anodic isoenzymes appearing, subsequently, at times which differ from one tissue to another. The isoenzyme pattern of the kidney of three-day-old animals is essentially that of the adult, whereas in the stomach, maturation of isoenzyme pattern takes three weeks, and even longer in the tongue (Markert and Ursprung, 1962). Changes in rat heart resemble those in the mouse tissue, with LD_5 as the initially most prominent isoenzyme, but with LD_1 taking its place soon after birth (Fine *et al.*, 1962). These observations have been interpreted as indicating a dependence on anaerboic glycolysis with the production of lactate for the production of energy during foetal life in both species.

The isoenzyme changes characteristic of rat and mouse tissues are not universally mirrored in other species. The predominant isoenzyme in embryonic chicken tissues is LD_1, and while there is little subsequent change during development of the heart, there is a gradual transfer of activity to the LD_4 and LD_5 isoenzymes during maturation of liver and skeletal muscle (Philip and Vesell, 1962). In higher animals, including man, both the main lactate dehydrogenase loci are expressed in embryonic tissues. Approximately equal amounts of the two types of subunits appear to be present in human tissues, so that LD_3 is the most prominent isoenzyme in extracts of both skeletal and heart muscle. The patterns of these tissues diverge during foetal development, so that the characteristic heart pattern, with its predominant anodal isoenzymes, and the skeletal muscle distribution weighted towards the LD_4 and LD_5 isoenzymes, are established by about the sixth month of intrauterine life (Dreyfus et al., 1962; Latner and Skillen, 1964; Werthamer et al., 1973). Smaller, quantitative changes in isoenzyme distribution may continue up to birth and into early post-natal life. An increased proportion of anodal lactate dehydrogenase isoenzymes has been noted in muscle tissue of normal subjects over 60 years of age, resembling the pattern seen in young children (Rosalki, 1968).

The third lactate dehydrogenase locus of mammals is only expressed in the primary spermatocyte and the homotetramer (LD_X or LD_C) composed of the polypeptide subunits controlled by it appears in testis and sperm only after puberty (Fig. 6.2). The absence from extracts of human testis of heteropolymeric isoenzymes containing C and H or M subunits, although such hybrid isoenzymes can be formed in vitro, suggests that the three loci are not active in the same cells.

Changes in other isoenzyme systems

Studies on X-linked isoenzymes such as glucose-6-phosphate dehydrogenase in the mouse have shown that both X chromosomes are active in the oocyte and in female embryos in the pre-implantation period (Brinster, 1979). Inactivation of one of the pair of X chromosomes postulated by the Lyon hypothesis must therefore take place after this stage.

Changes in distribution of several isoenzymes besides lactate dehydrogenase take place during the differentiation of human skeletal muscle.

The predominant isoenzyme of creatine kinase is the BB dimer, with smaller amounts of the MB hybrid isoenzyme, in samples obtained earlier than 12 weeks gestation. However, the MM dimer is the most prominent isoenzyme by the middle of pregnancy; subsequently, first the BB dimer

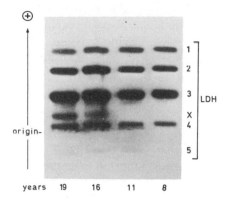

Fig. 6.2   *Isoenzymes of lactate dehydrogenase in human testis at various ages, showing the appearance of isoenzyme X in the mature tissue (From Edwards and Hopkinson, 1977. By permission of A.R. Liss, New York).*

then the MB hybrid disappears, so that the adult pattern in which virtually all the activity is in the form of MM-creatine kinase is established at or soon after the time of birth (Tzvetanova, 1971).

A progression during development towards the predominance of a single isoenzyme in the adult tissue is seen also in the case of *aldolase*. Although the A-type tetramer is already the major aldolase component in human foetal muscle after 20 weeks gestation, evidence for activity of the locus determining the C-type subunit is provided by the presence, in decreasing order of prominence, of A_3C, A_2C_2 and AC_3 hybrids (Edwards and Hopkinson, 1977). The hybrid isoenzymes are rapidly lost during further development, and muscle of the newborn, like that of the adult, contains essentially only A_4 tetramers. Marked developmental changes in the distribution of *enolase* isoenzymes take place in human skeletal muscle. The products of all three enolase loci are seen in the youngest foetal specimens, but the intermediate and fast isoenzymes, containing the subunits determined by the enolase-2 locus, are progressively lost (Edwards and Hopkinson, 1977). The persisting slow isoenzyme is itself heterogeneous and can be resolved into a form composed of products of the enolase-1 locus which also declines in activity during development, and a form determined by the enolase-3 locus which is the predominant isoenzyme of mature muscle (Pearce *et al.*, 1976).

As might be expected from the highly specialized nature of the metabolism of the liver, this tissue also shows characteristic changes in the patterns of several isoenzymes during development from foetus to adult.

The form of *hexokinase* with low affinity for glucose, hexokinase isoenzyme IV (also called glucokinase) is present in rat liver in foetal and early neonatal life at low activities only, but between the ages of 12 and 30 days its activity reaches levels six times greater than those of other hexokinase isoenzymes in this tissue (Walker, 1974). Hexokinase IV has not been demonstrated unequivocally in human liver, although its presence in placenta has been reported. The most marked change among other hexokinase isoenzymes in developing human liver is a decline in the activity of isoenzyme II, which is prominent in foetal liver (Edwards and Hopkinson, 1977). In early foetal development, all three *aldolase* isoenzymes, A, B and C, together with the various hybrid tetramers, can be detected in extracts of human liver (Hers and Joassin, 1961). However, at birth aldolase B is the predominant isoenzyme, as in adult liver. Striking changes in the distribution of isoenzymes of *alcohol dehydrogenase* occur in human liver during pre-natal development (Fig. 6.3). A single alcohol dehydrogenase isoenzyme is present in embryonic liver after about 10 weeks gestation, consisting of dimers of the α-polypeptide controlled by the ADH_1 locus. Increasing activity of the second ADH locus is shown by the appearance of β-polypeptides, seen first as hybrid dimers with α-polypeptides and then as $\beta\beta$-dimers, so that, at birth, three alcohol dehydrogenase isoenzymes ($\alpha\alpha$, $\alpha\beta$ and $\beta\beta$) are present. Continuing changes in the expression of these two loci increase the proportion of the $\beta\beta$-dimers relative to $\alpha\alpha$-dimers. Furthermore, expression of a third locus, ADH_3, leads to the appearance of γ-polypeptides and additional dimeric isoenzymes (Smith *et al.*, 1971a). The major part of the *pyruvate kinase* activity of human foetal liver is in the form of the M_2 isoenzyme, with only a small proportion contributed by the L isoenzyme, whereas in adult liver these proportions are reversed (Edwards and Hopkinson, 1977).

As in the case of the LD_X isoenzyme of lactate dehydrogenase, expression of the prostate-specific isoenzyme of *acid phosphatase* begins only with sexual maturity in the male.

Changes in the relative proportions of isoenzyme-producing organelles or cells

The changes in isoenzyme patterns during development discussed so far, result mainly from changes in the relative activities of multiple gene loci

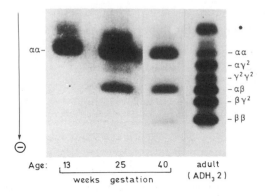

Fig. 6.3 *Changes in the expression of alcohol dehydrogenase isoenzymes determined by three genetic loci during development of human liver. The specimen of adult liver is from a subject of phenotype 2 at the ADH-3 locus which determines the γ-subunit. The nature and origin of the zone marked with an asterisk, which occurs in extracts of fresh liver, are unknown (From Edwards and Hopkinson, 1977. By permission of A.R. Liss, New York).*

within developing cells of a particular type, e.g. hepatocytes. Other alterations in the balance of isoenzymes within the whole organism may derive from changes in the number of subcellular organelles, or in the number or activity of cells which contain large amounts of a characteristic isoenzyme. As would be expected, both of the gene loci which determine pairs of mitochondrial and cytoplasmic isoenzymes are expressed in foetal tissues, although activities of mitochondrial isoenzymes are generally lower than those of the corresponding cytoplasmic isoenzymes throughout gestation. However, the relative proportions of cytoplasmic and mitochondrial forms often do not remain similar in all tissues as differentiation takes place; for example, the mitochondrial isoenzymes of isocitrate and malate dehydrogenases, aspartate aminotransferase and aconitase, show a relatively greater increase in activity than their cytoplasmic counterparts during the maturation of heart muscle compared with changes in other developing tissues (Edwards and Hopkinson, 1977). These differences are probably related, at least partly, to the relative abundance of mitochondria in the various tissues.

An example of the way in which changes in production of a characteristic enzyme form by a particular type of cell can alter the balance of forms in the whole organism is provided by the multiple forms of alkaline phosphatase. An increase in the number and activity of osteoblasts is responsible for mineralization of the skeleton between the early post-natal period and the beginning of the third decade of life. These cells progress through a life cycle in which the activity of alkaline phosphatase in them increases, then falls again as the cells become surrounded by bone matrix. The excess of alkaline phosphatase from the active osteoblasts enters the circulation, where its presence can be recognized by its characteristic properties and where it elevates the total serum alkaline phosphatase activity of young subjects above that of skeletally-mature adults. A similar, though distinguishable, alkaline phosphatase apparently originating in the liver, also contributes to the total activity of this enzyme in normal plasma, and the amount of this form in plasma shows a small, progressive increase with age (Whitaker *et al.*, 1977). The reason for this change is not known, but it may result from increased synthesis of the isoenzyme by hepatocytes in response to continuing exposure to factors which induce enzyme synthesis. Thus, the balance of the bone and liver alkaline phosphatases in plasma at a given age is determined by the relative activities of these two processes of enzyme production (Fig. 6.4).

Developmental changes in isoenzymes in plants

Changes in gene expression during the development of plants have been demonstrated through electrophoretic studies of their dependent isoenzymes, as in the case of multiple loci in animals and man (Scandalios, 1974). Among the isoenzymes of plants which have received particular attention in developmental studies are those of amylase and catalase in maize.

Two catalase isoenzymes are present in maize, determined by distinct genetic loci. Six allelic variants also occur at the Ct_1 locus without corresponding variation at the second, Ct_2, locus. Maize catalase is tetrameric and the homopolymers formed from Ct_1 and Ct_2 subunits have different physiochemical properties. Hybrid isoenzymes can be formed between the Ct_2 subunits and those determined by the Ct_1 alleles, and between the allelic Ct_1 subunits themselves, both *in vivo* and *in vitro*. The balance of activity contributed by each of the two catalase loci changes markedly during maturation of maize seeds, with the isoenzyme de-

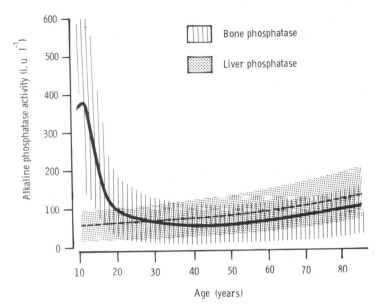

Fig. 6.4   *Changes in the relative activities of bone and liver alkaline phosphatases in human serum with age, determined by the quantitative heat inactivation technique of Moss and Whitby (1975).*

termined by the Ct_1 locus which is present in the post-pollination stage being replaced by the isoenzyme determined by the Ct_2 locus after germination. Hybrid isoenzymes are present during the transition from predominance of one homotetramer to that of the other (Scandalios, 1975).

Maize possesses distinct α- and β-amylases, coded at different loci as would be expected for these enzymes of distinct substrate specificities. Two codominant alleles are present at each locus. However, an interesting observation is that the alleles at the Amy-1 locus, which determine the structure of α-amylase isoenzymes, are not expressed synchronously during development (Chao and Scandalios, 1975).

Both soluble and mitochondrial isoenzymes of malate dehydrogenase, determined by distinct structural genes, are present in maize seedlings, as they are in animal tissues. Each of the plant isoenzymes shows considerable electrophoretic heterogeneity. As with the animal iso-enzymes, both maize isoenzymes are present early in development and the activities of both increase in germinating maize seeds (Yang, 1975).

Considerable attention has focused on the slime mould *Dictyostelium discoideum* in the search for simpler multicellular organisms in which the biochemistry of differentiation can be studied. The life cycle of this organism passes through clearly-defined stages with features reminiscent of both plant and animal life: an existence as single amoeboid cells is followed by a phase of aggregation into a multicellular tissue (the grex), which migrates before becoming stationary in a culmination phase which leads to the formation of a fruiting body. Cell differentiation appears at the culmination stage. The slime mould possesses distinct isoenzymic forms of several enzymes, the relative proportions of which change significantly at various stages of its life cycle (Loomis, 1975).

Two isoenzymes of *threonine deaminase* can be distinguished on the basis of several properties, including different sensitivities to inhibition by L-isoleucine and L-leucine. Isoenzyme 1 is sensitive to these inhibitors and therefore isoleucine synthesis can be regulated by feedback inhibition. Isoenzyme 1 activity is high during the amoebic phase but declines during aggregation and is low in later stages. Activity of the non-inhibited isoenzyme 2 increases after aggregation, probably representing an increased requirement for energy production by the catabolism of amino acids.

Changes in the relative activities of different isoenzymes of *β-glucosidase* and *alkaline phosphatase* are seen at the culmination stage. Glucosidase-1 of growing cells is replaced by isoenzyme 2, with a somewhat lower K_m for synthetic glucoside, as development begins. Changes in the proportions of electrophoretically-separable alkaline phosphatase isoenzymes also begin at the same time as those affecting the glucosidases. Two isoenzymes of *acetylglucosaminidase* are present in the organism. The activity of one remains low throughout the life cycle, whereas that of the second isoenzyme increases throughout aggregation and migration to reach a constant level by early culmination.

Developmental changes in multiple forms of enzymes due to post-genetic modifications

Changes in the distribution of multiple forms of enzymes which arise by post-translational modification of a single gene product are less frequent and less well-defined than those involving true isoenzymes. However, some trends can be discerned. The extent of post-translational modification seems to be less in foetal tissues than in adult specimens for a number of human enzymes, e.g. nucleoside phosphorylase, and the more rapid turnover of foetal cells with consequently reduced opportunities for post-translational modification of their constituent enzymes has been suggested as a possible explanation of this (Edwards and Hopkinson,

1977). The view that increased post-translational modification may be a consequence of ageing is supported by the observation that multiple enzyme forms which apparently arise in this way are rather common in erythrocytes, cells with a relatively long life-span and without the means to renew their enzymes by *de novo* synthesis. Furthermore, increases in anodal electrophoretic mobility have been reported when isoenzymes of glucose-6-phosphate dehydrogenase or aspartate aminotransferase prepared from old and young human erythrocytes are compared. These changes may involve oxidation of sulphydryl groups *in vivo*, as analogous changes seen on storage of haemolysates *in vitro* appear to do (Walter *et al.*, 1965).

Age-related changes in the characteristics of multiple forms of enzymes have also been observed in other human cells. Human fibroblasts have a limited life-span in tissue culture which is inversely related in length to the age of the donor. About 5% of the total glucose-6-phosphate dehydrogenase activity of young fibroblasts in culture is in a heat-labile form, but this proportion increases to 15–25% in older cells, which also contain an increased proportion of heat-labile 6-phosphogluconate dehydrogenase (Holliday and Tarrant, 1972). Fibroblasts from patients with Werner's syndrome, a familial condition of accelerated senescence and early death, contain a proportion of heat-labile glucose-6-phosphate dehydrogenase after only a short period of culture that is of the same order of magnitude as that found in aged cultures of normal fibroblasts (Holliday *et al.*, 1974). These changes may reflect the accumulation of errors in protein biosynthesis during ageing.

The age-dependent changes in production of liver and bone types of alkaline phosphatases by their parent cells have already been mentioned. It is probable that the tissue-specific differences between these forms do not derive from the existence of distinct structural genes; if, as seems more likely, the individual characteristics of each result from the operation of tissue-specific post-translational modifications, these processes appear to continue unchanged throughout life.

CHANGES ASSOCIATED WITH ABNORMAL DEVELOPMENT

Certain diseases, such as the progressive muscular dystrophies, appear to involve a failure of the affected tissues to mature normally or to maintain a normal state of maturity. Cancer cells show a progressive loss of the attributes of structure and metabolism of the healthy cells from which they arise. It is to be expected, therefore, that when the mature and fully-differentiated tissue displays a characteristic complement of isoenzymes, this pattern may be lost or modified if normal differentiation is arrested or reversed, and many examples have been reported of changes in the distribution of multiple forms of enzymes accompanying such processes.

A resemblance between dystrophic human skeletal muscle and foetal muscle in their isoenzyme patterns has been demonstrated in many studies, though not in all. The distributions of isoenzymes of aldolase, lactate dehydrogenase and creatine kinase in the muscles of patients with progressive muscular dystrophy have been found to be similar to those seen in the earlier stages of development of foetal muscle (Dreyfus *et al.*, 1962; Wieme and Herpol, 1962; Schapira *et al.*, 1968; Tzvetanova, 1971). Similar though generally less pronounced changes have also been reported in skeletal muscle from patients with less severe inherited or acquired myopathies, and accompanying atrophy of disuse and ageing. The isoenzyme abnormalities seen in dystrophic muscle have been interpreted as a failure to maintain, or even to reach, a normal degree of differentiation. The rather similar changes of ageing or disuse may also represent dedifferentiation, with alterations in the function of regulatory mechanisms which normally determine the relative proportions of the various isoenzymes in the differentiated tissue. Isoenzyme patterns may also show some tendency to approach foetal distributions in regenerating tissues. This may result from relaxation or modification of control systems in rapidly dividing cells and may account for some of the isoenzyme changes seen, for example, in muscle in acute polymyositis.

Hypertrophy of cardiac muscle with hypoxia produces a relative increase in the proportion of M-subunits of lactate dehydrogenase. This has been interpreted as consistent with a trend towards anaerobic metabolism of cells as a consequence of an increased capillary-to-cell diffusion distance. However, an alternative possibility is that the M-subunits are present in fibrous tissue, which constitutes a greater proportion of the tissue of the hypertrophic heart than is the case in the normal organ (Revis *et al.*, 1977).

Changes associated with malignant transformation

A re-emergence of foetal patterns of isoenzyme distribution is a feature of malignant transformation in many tissues. This phenomenon was first studied extensively in the case of lactate dehydrogenase isoenzymes. Malignant tumours in general show a significant shift in the balance of isoenzymes towards the electrophoretically more cathodal forms, LD_4 and

LD_5 (Goldman *et al.*, 1964). The decline in activity of the LD_1 and LD_2 isoenzymes results in patterns which are reminiscent of those occurring in embryonic tissues. Tumours of prostate, cervix, breast, brain, stomach, colon, rectum, bronchus and lymph nodes are among those which show this transformation. In contrast, comparatively benign gliomas show a relative increase in anionic isoenzymes. A relative increase in the proportion of cathodal isoenzymes of lactate dehydrogenase has also been observed in tissues adjacent to malignant tumours, e.g. of the colon, although the cells in these regions are morphologically normal (Langvad, 1968). Similar changes in the distribution of lactate dehydrogenase isoenzymes are seen in pre-invasive carcinoma of the cervix uteri (Latner *et al.*, 1966), and may represent an early manifestation of metabolic change in cells which subsequently become malignant.

The pronounced glycolytic activity of malignant tumours first noted by Warburg and his collaborators in the early 1920s and the apparent suitability of the more cathodal isoenzymes of lactate dehydrogenase for this type of metabolism have been linked by the suggestion that a shift to predominantly glycolytic metabolism may be a fundamental aspect of malignant change, and that changes in lactate dehydrogenase isoenzyme patterns are an essential accompaniment or even a causative factor of such change. However, the subsequent accumulation of evidence that other isoenzyme patterns undergo changes in malignant tumours and that some of these changes do not appear to represent specific adaptations to glycolytic metabolism has made this view less attractive than theories which view the isoenzyme changes as part of a general loss of differentiation. Loss of a characteristic pattern of lactate dehydrogenase isoenzymes with reversion to a more primitive distribution has been reported for several cell lines maintained in culture.

As well as lactate dehydrogenase, the isoenzyme patterns of aldolase, pyruvate kinase and hexosaminidase have been shown to undergo a change towards foetal-like patterns in hepatoma (Schapira *et al.*, 1975). These changes were also present, but to a less marked extent, in regenerating liver. A further eleven isoenzyme systems which differ in neoplastic cells from their counterparts in normal tissues have been listed by Criss (1971), including examples such as malate dehydrogenase and aspartate aminotransferase in which the balance between cytoplasmic and mitochondrial isoenzymes is altered.

In 1968, an alkaline phosphatase was identified in the serum of a patient with metastatic squamous cell carcinoma of the lung which was identical with the alkaline phosphatase of normal placenta with respect to inhibition by L-phenylalanine, heat stability and alteration of electrophoretic mobility by digestion with neuraminidase. The newly-discovered isoenzyme, which was termed the Regan isoenzyme after the patient in whom it was discovered, was also precipitated by an antiserum to human placental alkaline phosphatase (Fishman *et al.*, 1968) (Fig. 3.5a). Further evidence of similarity between placental alkaline phosphatase and the Regan isoenzyme was provided by comparison of the purified isoenzyme from normal placenta with a preparation of alkaline phosphatase from liver metastases of a giant-cell carcinoma of the lung (Greene and Sussman, 1973). The two enzymes had identical subunit molecular weight and their isoelectric points were closely similar. The placental and tumour-derived alkaline phosphatases had identical N-terminal amino acid sequences over the last four residues, and this sequence was different from that of the liver isoenzyme. Two-dimensional peptide maps of partially digested alkaline phosphatases of tumour and placenta showed slight differences but these differences were close to the limits of experimental variation. These results supported the view that these two isoenzymes were products of the same structural gene with some small differences resulting from post-genetic modification.

The Regan isoenzyme originally described accounted for about half of the elevated alkaline phosphatase activity of the patient's serum and it was present also in large quantities in extracts of the tumour tissue. Since its first description, the Regan isoenzyme has been detected in the sera of patients with many types of malignant disease, and also in some patients with non-malignant diseases. An incidence of the isoenzyme of 3–15 % in cancer patients has been estimated, but this varies with the sensitivity of the methods used for its detection.

Normal placental alkaline phosphatase isoenzymes display various rates of migration on electrophoresis depending on their phenotype. Regan isoenzymes show a similar variation in electrophoretic mobility from one patient to another, and their mobilities can be related to those of the known normal phenotypes. However, an enzyme form with a mobility slower than the common isoenzymes of placental alkaline phosphatase is encountered fairly frequently amongst cancer patients who exhibit placental phosphatase-like isoenzymes, occurring in about half of such patients in one series. The mobility of this isoenzyme corresponds to that of one of the rarer variants of placental alkaline phosphatase, known as the D variant. The D variant of placental alkaline phosphatase is inhibited by L-leucine, as well as by L-phenylalanine, and this property distinguishes the D variant from the other common isoenzymes of placental alkaline phosphatase. Inhibition by L-leucine had previously been reported as a property of an unusual form of alkaline phosphatase found in the serum of a patient with pleuritis carcinomatosa and called by its discoverers the Nagao isoenzyme (Nakayama *et al.*, 1970). A similar isoenzyme was subsequently isolated from the primary tumour of an adenocarcinoma of the pancreas and from secondary deposits from the liver in adenocarcinoma of the bile duct (Jacoby and Bagshawe, 1971). Although it seems that the Nagao isoenzyme and the D variant of placental alkaline phosphatase are very similar, differences between the two have been demonstrated in their degrees of inhibition by L-leucine-containing peptides (Doellgast and Fishman, 1977).

Yet another alkaline phosphatase resembling the placental isoenzyme in many respects has been identified in hepatoma tissue (Warnock and Reisman, 1969). This variant had a greater mobility towards the anode on starch gel electrophoresis than normal liver phosphatase or any allelo zyme of placental alkaline phosphatase. It was also less stable than normal placental phosphatase when heated at 65°C, but much more stable in this respect than non-placental alkaline phosphatases such as the liver isoenzyme. Subsequently, the alkaline phosphatase of hepatoma was shown to be immunologically identical to the normal placental isoenzyme, but these two enzyme forms could be differentiated by inhibition with EDTA (Higashino *et al.*, 1972).

Excessive production, by neoplastic tissues, of variants of alkaline phosphatase unlike the placental isoenzyme has also been observed (Timperley, 1968; Timperley *et al.*, 1971; Romslo *et al.*, 1971). These enzymes are heat-labile and are usually not markedly inhibited by L-phenylalanine. In these respects they resemble the alkaline phosphatases of liver and bone, and the electrophoretic mobilities of the phosphatase derived from the tumour tissue may also be similar to those of the normal liver or bone isoenzymes. However, it is not certain that these tumour enzymes are completely identical to any normal tissue isoenzyme. A variant alkaline phosphatase has been found at high levels of activity in both primary and secondary tumour tissue from a patient with the kidney tumour, hypernephroma. The phosphatase variant was also present in significant amounts in the serum. In this case the abnormal enzyme was rather similar to bone phosphatase but unlike normal kidney phosphatase in electrophoretic mobility. It was slightly less stable at 56°C than normal kidney phosphatase but more stable than the bone enzyme (Whitaker *et al.*, 1982).

The appearance in neoplastic tissues of forms of alkaline phosphatase which appear to be identical with the normal placental isoenzyme seems to be an analogous phenomenon to the ectopic production of other proteins or polypeptides by malignant cells. Indeed, in at least one case of mediastinal choriocarcinoma, parallel changes in levels of carcinoembryonic antigen, chorionic gonadotrophin and a Regan isoenzyme of alkaline phosphatase were observed (Belliveau *et al.*, 1973). Malignant transformation thus appears to result in depression or reactivation of the genes which determine the structure of these proteins in some tissues. However, some enzyme variants discovered in malignant tissues appear to be altered forms of enzymes which are normally expressed in those tissues. For example, altered forms of glucose-6-phosphate dehydrogenase have been detected in granulocytes from patients with certain forms of leukaemia and myeloproliferative disorders. The changes in the properties of the enzyme included decreased isoelectric points and decreased affinity for glucose-6-phosphate, as well as reduced molecular activity. Some of the differences between the normal and modified forms of glucose-6-phosphate dehydrogenase could be reproduced by incubating extracts of malignant cells *in vitro*, suggesting that they are due to post-translational modification of the enzyme molecule in the malignant cells (Kahn *et al.*, 1975). A further possibility is that a somatic mutation of an enzyme-determining gene in malignant cells may alter the structure of its product. This may account for the appearance of the variant alkaline phosphatase in hypernephroma already mentioned, as well as for additional or altered forms of lactate dehydrogenase which have been observed in the sera of cancer patients (Beautyman, 1962). Hybridization of abnormal lactate dehydrogenase subunits with the normal monomers may give rise to variant tetramers which appear as enzyme zones of abnormal mobility on electrophoresis of serum or extracts of tumour tissue (Fujimoto *et al.*, 1968).

Both re-expression of a latent gene and its modification by mutation may take place in malignant cells, possibly accompanied also by post-translational modification of its product, and a combination of these factors may account for the observation that placental alkaline phosphatases reappearing in cancer cells are not always identical with the normal isoenzyme.

Increased amounts of alkaline phosphatase of placental type are produced by certain cultured cell lines fron non-placental human tissues, e.g. by some sublines of HeLa cells (Fishman *et al.*, 1968; Elson and Cox, 1969). Alkaline phosphatases which are of non-placental type are also produced in considerable amounts in certain cultured cells, sometimes occurring together with placental phosphatase-like isoenzymes and sometimes alone. While in some cell lines the alkaline phosphatase may be very similar to either normal placental- or non-placental alkaline phosphatases, respectively, even to the extent of apparent immunological identity, differences in detailed properties are often found. For example, the alkaline phosphatase of cultured cells may appear as a series of multiple zones on electrophoresis which do not coincide exactly with zones of the enzyme seen in extracts of normal tissues. Differences in response to inhibition may also exist between the alkaline phosphatases of cultured cell lines and their most closely-similar counterparts from normal human tissues.

Thus, as with the alkaline phosphatases produced ectopically by neoplastic tissues *in vivo*, the alkaline phosphatases occurring in cells in culture frequently appear to have undergone structural and functional modifications to varying extents, either post-genetically or as a result of alterations in the genes which determine their structures.

Selective induction of isoenzymes

Many examples are known of induction of enzyme activity, by which exposure to certain agents such as drugs increases the activity of a particular enzyme by stimulating increased synthesis of it. When an inducing agent acts selectively on a particular isoenzyme, the balance of isoenzymes making up the total enzymic activity in question is changed. Several different kinds of stimuli have been shown to selectively increase the alkaline phosphatase activity of various tissues. These include dietary phosphate, which increases rat kidney alkaline phosphatase; vitamin D, which induces intestinal alkaline phosphatase in chicks; and cortisone or similar substances, which produce an increased activity of alkaline phosphatase in human leukocytes, mouse duodenum and in some HeLa cell lines. Catecholamines or their synthetic analogue isoprenaline cause an increased alkaline phosphatase activity localized specifically to the right atrium of rat heart. However, one of the most interesting of such phenomena from the clinical point of view is the increase in alkaline phosphatase activity in the liver which follows occlusion of the bile duct.

For many years, the increased alkaline phosphatase in serum seen in pathological or experimental obstruction of the bile duct was attributed to retention of alkaline phosphatases from extra-hepatic sources, chiefly bone, which would normally be excreted in the bile. However, isoenzyme characterization of serum and tissue alkaline phosphatases revealed that the properties of the enzyme in serum in cholestasis resembled liver alkaline phosphatase kinetically and electrophoretically (Moss *et al.*, 1961a; Hodson *et al.*, 1962), rather than the isoenzyme from bone. Direct evidence that the liver was the source of increased alkaline phosphatase activity in biliary obstruction was provided by experiments which showed that occlusion of the bile duct from a single lobe of dog liver caused increased activities of serum and liver alkaline phosphatases in the absence of hyperbilirubinaemia (Polin *et al.*, 1962), and increased tissue and perfusate alkaline phosphatase levels resulted from bile duct ligation in isolated, perfused cat liver (Sebesta *et al.*, 1964). The rise in the activities of rat liver and serum alkaline phosphatases which typically follows ligation of the bile duct was prevented by previous administration of cycloheximide or actinomycin D, indicating the role of protein synthesis in the increase in enzyme activity in cholestasis (Kaplan and Righetti, 1970). Experiments in which cells of different types have been isolated from normal and cholestatic rat livers have shown that the rapid increase in alkaline phosphatase activity following biliary occlusion takes place largely in parenchymal cells (Wootton *et al.*, 1977).

The most probable explanation for the increased alkaline phosphatase activity in cholestasis is *de novo* synthesis of a pre-existing alkaline phosphatase isoenzyme, and this is supported by the general similarity of properties of alkaline phosphatases prepared from livers of normal and cholestatic rats (Righetti and Kaplan, 1974). An increase in the amount of alkaline phosphatase protein in the livers of rats following ligation of the bile duct was demonstrated by immunochemical titration (Schlaeger, 1975). The nature of the stimulus which operates at the cellular or molecular level to increase the production of alkaline phosphatase in the liver is not known.

The enhancing effect of cortisone or its analogues on the alkaline phosphatase activity of certain cells, such as some HeLa cell lines, differs in some respects from the effects of other inducers of alkaline phosphatase. In the case of HeLa cells, the increased activity of alkaline phosphatase produced by prednisolone is not accompanied by an increase in the amount of enzyme protein measured by an immunochemical technique (Cox *et al.*, 1971). The alkaline phosphatase of HeLa cells is a phosphoprotein, and, after induction with prednisolone, the enzyme molecule contains approximately half as many inorganic phosphate residues as the basal form. This modification apparently alters the binding forces involving the zinc atom at the active centre of the phosphatase, producing a catalytically more efficient molecule (Bazzell *et al.*, 1976).

However, if synthesis of RNA and protein is prevented by appropriate inhibitors, the increase in alkaline phosphatase activity in response to treatment with prednisolone does not take place. The process is therefore thought to involve the synthesis of a modifier, the effect of which is to interact with the alkaline phosphatase to increase its molecular activity.

Prednisolone has been shown to have a selective effect on particular isoenzymes in HeLa cell lines which exhibit more than one form of the enzyme on electrophoresis of extracts of the cells. Thus, treatment of cultures of some cell lines with prednisolone produces an enhancement of some enzyme zones or even the appearance of new zones, and a reduction in the intensity of others. The enhancing effect of prednisolone appears to be more pronounced for enzyme zones of the placental phosphatase-like, or Regan, type (Singer and Fishman, 1975).

As in normal maturation, an increase in the number or activity of particular cells which are rich in a specific form of an enzyme as a result of disease will alter the balance of activity of the multiple forms of that enzyme in the whole organism. For example, any condition which stimulates osteoblastic activity will cause an increase in the contribution of the osteoblastic variant of alkaline phosphatase to the total alkaline phosphatase activity. Osteoblastic activity is increased in pathological conditions such as rickets and osteomalacia, Paget's disease of bone, hyperparathyroidism with skeletal involvement, osteogenic sarcoma, or neoplastic metastases in bone. The increased production of the osteoblastic variant is reflected in a change in the pattern of multiple forms of alkaline phosphatase in serum. Similarly, when a neoplasm produces a specific enzyme variant, or continues to express a characteristic isoenzyme, it and its secondary deposits constitute additional isoenzyme sources, the presence of which may be reflected in the isoenzyme pattern of the serum. Thus, increased prostatic acid phosphatase in serum is associated with metastatic carcinoma of the prostate, and a considerable proportion of advanced malignancies have increased serum lactate dehydrogenase activities, often due to the electrophoretically-slow isoenzymes typical of tumour cells.

7 Multiple Forms of Enzymes in Diagnostic Enzymology

Almost all the enzymes of the human body function within the cells in which they are formed, relatively few, such as the digestive enzymes and the enzymes of the blood clotting system, being secreted into the extracellular fluids. Although small samples of almost any organ can now safely be obtained by biopsy techniques, chemical analysis of these samples, including the measurement of enzyme activities, has so far found little diagnostic application. In contrast, studies of the characteristics and activities of enzymes in readily available samples of body fluids, usually serum, are widely applied in the investigation of disease, in an attempt to infer the nature and extent of pathological changes within the cells and tissues. The enzymes chosen for study in diagnostic enzymology need not necessarily be primarily involved in the process of disease provided that a correlation can be established between their extracellular levels and particular pathological processes.

Although much useful information can be gained simply from measurements of the total activities of individual enzymes in serum, most of the enzymes which have been shown to be of diagnostic significance occur widely throughout the body, so that such measurements do not by themselves provide information as to the identity of the affected tissue in the absence of other evidence. The relative activities of several enzymes in a serum sample may reflect their relative activities in the tissue from which the enzymes originate. However, differential rates of release of enzymes from cells, or differential rates of removal from the circulation, may modify the pattern of enzyme activities in serum, while disease may be accompanied by changes in the rates of enzyme synthesis in the affected tissue.

The existence of multiple enzyme forms with tissue-specific patterns of

159

:ion offers an alternative means of identifying affected organs by tests. The effectiveness of this approach requires that, as well as ──···─ₒ ─ non-uniform distribution, the multiple forms should retain their identifying characteristics after they are released from their cells of origin. For isoenzyme studies to be useful in clinical practice, these characteristics should be sufficiently distinctive to enable them to be demonstrated reliably by relatively simple techniques.

The total activity of a particular enzyme in serum is the sum of the individual activities of those of its multiple forms which may be present. The levels of enzyme activity in healthy individuals vary considerably from person to person, and may also show quite large variations in one person as a result of such physiological causes as changes in diet or amount of exercise, or in rate of growth or hormonal status. The need to take into account these possibilities when interpreting the results of enzyme analyses made for diagnostic purposes implies that rather wide ranges of serum activity must be regarded as potentially normal, thus reducing the sensitivity of the tests in the detection of small abnormalities. These uncertainties of interpretation can be greatly reduced by measuring specifically the activity of an isoenzyme (or other multiple enzyme form) derived from a single tissue; i.e. it may be possible to detect a significant change in the activity of a specific isoenzyme in serum even though the total enzymic activity remains within normal limits.

The extent to which these two advantages, of greater organ-specificity and increased sensitivity, can be realized in diagnosis is closely related to the analytical methods employed and many improvements in methodology have resulted from the needs of diagnostic enzymology. Indeed, much of the impetus for the study of multiple forms of enzymes in general has originated in a desire to extend the value of enzyme tests in diagnosis.

LACTATE DEHYDROGENASE ISOENZYMES

The interest of clinical enzymologists in the increased possibilities of organ-specific diagnosis offered by the existence of isoenzymes was first aroused by the description of the specific patterns of distribution of lactate dehydrogenase isoenzymes in the late 1950s. Hopes were expressed of achieving diagnostic dissection by isoenzymes, in which a specific pattern of lactate dehydrogenase isoenzymes in serum could be associated with each of a wide range of organs or tissues. These early hopes have had to be relinquished with the realization that the patterns of distribution of the isoenzymes are limited to three, each shared by several tissues. Nevertheless, lactate dehydrogenase isoenzymes have formed a model for later applications and have retained a more limited but useful role in diagnostic enzymology.

Determinations of lactate dehydrogenase isoenzyme patterns in serum offer advantages of sensitivity in the detection of minimal cardiac muscle damage, and of organ-specificity in distinguishing between cardiac and hepatic damage in certain circumstances. Inspection of the patterns of lactate dehydrogenase zones separated by electrophoresis and stained for enzyme activity is usually the only procedure necessary when the purpose of analysis is to distinguish between heart and liver as potential sources of an increased activity of the enzyme in serum (Fig. 7.1). The need

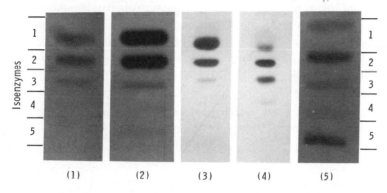

Fig. 7.1   *Patterns of lactate dehydrogenase zones in normal serum (1), in serum after myocardial infarction (2,3) showing increase in isoenzymes 1 and 2, in lung cancer (4) showing relative increase in isoenzyme 3, and in infective hepatitis (5) showing appearance of isoenzyme 5. Separation was by electrophoresis on agar-gel (1, 2 and 5) and cellulose-acetate (3, 4). The anode is at the top.*

to draw this distinction may arise in patients in the post-myocardial infarction period, when a second rise in serum enzyme activities may result from reinfarction or from congestive liver damage. The appearance of the LD_5 isoenzyme in serum in these circumstances indicates hepatocellular damage. However, this particular application of isoenzyme techniques has been rendered less important in recent years by the development of other tests of cardiac or hepatic tissue damage. For example, reinfarction causes a second rise in serum creatine kinase activity whereas liver damage does not, and the activity of isocitrate dehydrogenase in serum rises as a result of hepatocellular damage but not as a result of myocardial infarction. Nevertheless, isoenzyme studies of lactate dehydrogenase or creatine kinase, or both, may be necessary to locate the site of tissue damage in certain circumstances, e.g. in the first few days after surgical operations when the total activities of both these enzymes, as well as of other enzymic indicators of tissue damage, are usually considerably elevated.

The pattern of lactate dehydrogenase isoenzymes in serum remains abnormal following acute myocardial infarction for several days, after total lactate dehydrogenase activity and the levels of creatine kinase and aspartate aminotransferase have returned to normal. Thus, demonstration of an abnormal distribution of lactate dehydrogenase activity among the isoenzymes present in serum is the most sensitive enzymic indicator of myocardial infarction when investigation has been delayed for more than two days from the onset of symptoms. This is probably the most useful diagnostic application of lactate dehydrogenase isoenzymes; however, quantitative methods are required fully to exploit its potential value.

Increased activity with α-oxobutyrate as substrate instead of pyruvate also provides a quantitative indication of a shift in the balance of isoenzymes towards the more anodal forms, as does an increased proportion of heat-resistant lactate dehydrogenase activity. However, these methods are insensitive to small changes in the relative proportions of LD_1 and LD_2 and so cannot detect a reversal of the normal ratio, in which activity of LD_2 exceeds that of LD_1. A reversed ratio of these two isoenzymes, which is the most sensitive indicator of an altered lactate dehydrogenase isoenzyme pattern of cardiac origin, can be demonstrated by densitometry of electrophoretic patterns stained to show zones of lactate dehydrogenase activity.

About one half of patients with pulmonary emboli have elevated serum lactate dehydrogenase activities and, in some, the isoenzyme pattern indicates that the liver is the source of the increased activity, allowing myocardial infarction to be excluded. However, LD_1 may be raised in pulmonary embolism, probably as a result of release of anodal isoenzymes from erythrocytes as a result of intravascular haemolysis. Possible interference by red cell isoenzymes, released either by intra-vascular haemolysis (e.g. in patients with artificial heart valves) or as a result of haemolysis during sample collection, is an important disadvantage in the use of lactate dehydrogenase isoenzymes in the diagnosis of myocardial damage. Very high levels of the more anodal isoenzymes of lactate dehydrogenase are found in sera from cases of untreated megaloblastic anaemias. The total enzyme activity and isoenzyme distribution return rapidly to normal when specific therapy with vitamin B_{12} or folic acid is instituted. The origin of the excessive amounts of LD_1 and LD_2

in megaloblastic anaemia is presumably the immature and abnormal erythrocyte precursors which are rapidly broken down in the bone marrow (Emerson *et al.*, 1967). Megaloblastic marrow shows a shift in lactate dehydrogenase isoenzyme pattern towards the anodal fractions compared with normal marrow.

In spite of the importance of skeletal muscle as a source of lactate dehydrogenase, studies of isoenzymes of lactate dehydrogenase in serum have proved to be of relatively little value in the diagnosis of muscle disease. In progressive muscular dystrophy, in which the highest levels of lactate dehydrogenase of muscle origin are reached, the isoenzyme pattern in serum frequently shows an increase in the more anodal isoenzymes, LD_1 and LD_2. The serum isoenzyme pattern in this condition presumably reflects the abnormal pattern in dystrophic muscle, in which the usual preponderance of cathodal isoenzymes is not present. The increase in anodal isoenzyes in serum which has been observed also in polymyositis is more difficult to explain, but selective enzyme release from muscles with a high proportion of red fibres containing predominantly anodal iso-enzymes may be the origin of the pattern seen in serum. Regeneration of muscle may also be a factor. However, the balance of lactate de-hydrogenase isoenzymes in serum may be shifted towards the electrophoretically-slower forms in both muscular dystrophies and other muscle diseases. This may represent a loss of LD_5 from muscle fibres accompanying dedifferentiation in the earlier stages of disease.

An increased activity of lactate dehydrogenase in serum is found in about one third of patients with malignant disease, the elevations being greater and more frequent when the disease is widespread, as for example in carcinomatosis or Hodgkin's lymphoma. The pattern of lactate dehydrogenase isoenzymes in the serum is variable. An increase in LD_3 is often observed and this may reflect the displacement of the pattern towards the more cathodal isoenzymes which generally takes place in the tumour tissue itself (Fig. 7.1). The cathodal shift of the isoenzyme pattern is usually more pronounced in malignant effusions than in serum. However, the serum isoenzyme pattern may also be influenced by destruction of normal tissues surrounding the tumour; e.g. when the liver has been invaded, LD_5 may appear in the serum. Increased activity of the anodal isoenzymes, LD_1 and LD_2, has been reported in the sera of patients with certain tumours, including disgerminoma of the ovary, seminoma and teratoma of the testis (Zondag and Klein, 1968). Additional or altered forms of lactate dehydrogenase may appear in the sera of cancer patients. These presumably originate from somatic mutation of lactate dehydrogenase subunits in the malignant cells. Hybridization of the altered subunits with normal subunits may give rise to variant tetramers of lactate dehydrogenase which appear as enzyme zones of abnormal mobility on electrophoresis of serum or extracts of tumour tissue (Fujimoto *et al.*, 1968).

Zones of lactate dehydrogenase in serum with atypical electrophoretic mobilities may result from association of lactate dehydrogenase molecules with IgA or IgG immunoglobulins. These zones are usually diffuse and migrate between LD_3 and LD_4, or LD_4 and LD_5. No pathological features appear to be common to the various patients in whom macromolecular complexes of lactate dehydrogenase have been detected; these zones do not, therefore, appear to be of diagnostic significance. The total serum lactate dehydrogenase activity may be elevated, perhaps because of interference with the normal processes of elimination of the enzyme from the circulation.

CREATINE KINASE ISOENZYMES

The value of determining the distribution of creatine kinase isoenzymes in serum lies principally in the detection of damage to the myocardium. This tissue contains a considerable amount of the hybrid MB isoenzyme of creatine kinase and, while cardiac muscle is not the only tissue in which the MB isoenzyme occurs, the myocardium is effectively the only tissue from which the hybrid isoenzyme enters the circulation in significant quantities. Thus, demonstration of an increased activity of MB-creatine kinase in serum is the most specific enzymic indicator of myocardial damage, such as myocardial infarction. Also, since the activity of the MB isoenzyme is very low in normal serum but increases on average more than tenfold after myocardial infarction, quantitative measurement of the MB isoenzyme is potentially a very sensitive test, capable of revealing minor damage to cardiac muscle cells (Figs. 7.2 and 7.3).

The normal level of activity of creatine kinase in serum derives apparently from skeletal muscle and takes the form entirely, or almost entirely, of the MM isoenzyme. After myocardial infarction or other damage to the myocardium, further MM isoenzyme from cardiac cells is released together with a smaller amount of the MB isoenzyme. The realization of the diagnostic value of MB creatine kinase in the investigation of heart diseases is therefore critically dependent on the use of methods which are able to detect and measure small changes in the activity of the MB isoenzyme in the presence of a much greater activity of MM-creatine kinase. A considerable number of methods have been devised which have attempted to combine the necessary specificity and

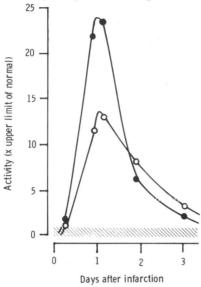

Fig. 7.2   *Time course of changes in total creatine kinase activity (open symbols) and activity of creatine kinase MB isoenzyme (solid symbols) in serum after an acute myocardial infarction, expressed as multiples of the appropriate upper reference limits in each case.*

sensitivity for the MB isoenzyme with the convenience and rapidity of use necessary for the analysis of large numbers of clinical specimens.

The three isoenzymes of creatine kinase differ sufficiently in net molecular charge to enable them to be separated completely by zone electrophoresis on various supporting media (Fig. 7.4). Agarose, agar and polyacrylamide gels and cellulose acetate have all been used for this purpose. Several methods have been applied to provide quantitative estimates of the relative activities of the separated isoenzyme zones, including densitometric or fluorimetric scanning of the zones after reduction of $NADP^+$ to NADPH by enzymic reactions coupled to the primary creatine kinase reaction (Somer and Konttinen, 1972). Visual assessment of the intensity of coloured or fluorescent MB-isoenzyme zones after electrophoretic separation has also been used for diagnostic purposes. However, changes in activity of the MB isoenzyme, such as those which occur in reinfarction, are not easily detected in this way. Furthermore, small amounts of the MB isoenzyme may just be detectable in normal sera by some techniques, making the method unreliable for the detection of minimal cardiac injury.

Several authors have preferred ion-exchange chromatography to electrophoresis as a way of separating the isoenzymes of creatine kinase, with continuous gradient or stepwise elution from DEAE-Sephadex or DEAE-cellulose (Mercer, 1974). Stepwise elution procedures in ion-

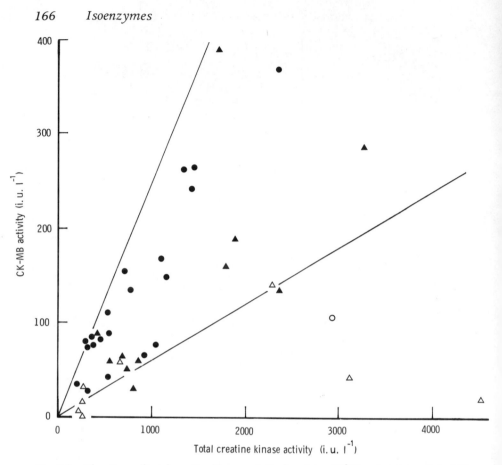

Fig. 7.3   *Elevations of total creatine kinase activity (i.u. 1-¹ at 37° C) and activity of the MB isoenzyme after recent acute myocardial infarction (solid symbols) and in various diseases of skeletal muscle (open symbols). The MB isoenzyme contributes between approximately 10 and 20% (lower and upper diagonal lines) of the total activity after uncomplicated myocardial infarction, whereas the proportion is generally below 10% in muscle disease. Circles indicate results obtained by electrophoresis and triangles values by immunoinhibition. Upper reference limits are 200 i.u. 1-¹ and 12 i.u. 1-¹ for total and MB activities, respectively.*

exchange chromatography suffer from the disadvantage that the first step may not elute the first isoenzyme completely, so that the residual fraction of it is washed off at the second step and is measured with the second isoenzyme. The MM isoenzyme of creatine kinase has the lowest net charge at alkaline pH and is eluted first from ion-exchange columns; since this isoenzyme is always present in serum samples in excess of the MB isoenzyme, even a slight carryover of MM isoenzyme into the second elution stage introduces a considerable error into the estimation of the MB

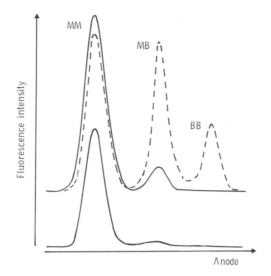

Fig. 7.4   *Separation of isoenzymes of creatine kinase is serum by electrophoresis on cellulose acetate, followed by fluorimetric scanning. The MB isoenzyme accounts for 315 of the total activity of 2335 i.u. 1-¹ in the serum from a case of myocardial infarction (upper trace). A small proportion of this isoenzyme is present also in the serum of a patient with Duchenne muscular dystrophy, making up 100 of the 2935 i.u. 1-¹ total activity (lower trace). The separation of a control serum containing all three isoenzymes of creatine kinase is also shown (broken line).*

isoenzyme. Nevertheless, stepwise elution from very small ion-exchange columns is widely used to separate the isoenzymes of creatine kinase. Other methods of separating creatine kinase isoenzymes include high-pressure liquid chromatography (Kudirka *et al.*, 1975) and batchwise adsorption on resin or glass bead ion-exchangers (Morin, 1976; Henry *et al.*, 1975).

The MM and MB isoenzymes of creatine kinase differ to some extent in their catalytic properties and, although they are not very marked, these differences have been made the basis of methods for measuring the activities of the individual isoenzymes in serum. Differential substrate inhibition has been used in this way (Witteveen *et al.*, 1974). This particular method makes use of the luciferase reaction, in which light is generated in the presence of ATP, as a sensitive way of measuring creatine kinase activity. The distinctive antigenic properties of the M and B subunits of creatine kinase have formed the basis of several immuno-chemical methods for determining the isoenzymes, by precipitation or inhibition, and a radioimmunoassay has also been described (Jockers-Wretou and Pfleiderer, 1975; Gerhardt *et al.*, 1977; Roberts *et al.*, 1976).

Immunochemical methods, especially the use of inhibitory antisera, seem likely to offer the combination of high sensitivity with applicability to automatic analysers which is desirable for diagnostic applications. However, all B subunits are measured in immuno-inhibition methods with a single antiserum, whether they are present as MB or BB dimers. A growing number of reports of the appearance of BB creatine kinase in the serum in a variety of non-infarctive conditions indicate the dangers of interpreting all elevated activities of B subunit as being due to the MB isoenzyme. Identification of the isoenzyme zones by electrophoresis may be necessary in doubtful cases. The lack of specificity of single-antibody assays can be overcome by two-site immunoradiometric methods (Willson *et al.*, 1981; Usategai-Gomez *et al.*, 1981).

The considerable effort which has been expended on the development and improvement of methods for the assay of creatine kinase isoenzymes in serum is a measure of the diagnostic value which is attached to these estimations. It illustrates also the crucial part played by method development in translating the potential advantages of isoenzyme studies into reliable diagnostic tests. Numerous studies have demonstrated the advantage of sensitivity and specificity to be gained in the investigation of suspected myocardial infarction provided that satisfactory methods of measuring the activity of the MB isoenzyme are available (Konttinen and Somer, 1973; Wagner *et al.*, 1973; Smith *et al.*, 1976). The increased specificity of MB isoenzyme measurements for myocardial damage is particularly valuable when creatine kinase activity in serum is already elevated due to leakage of the enzyme from skeletal muscle, e.g. in the period after surgical operations. There is an appreciable risk of myocardial infarction in this period, especially after cardiac surgery such as coronary artery bypass grafting. Although total serum creatine kinase activity is elevated two- to five-fold in the immediate post-operative period, depending on the duration and extent of the operation, and remains abnormal for about five days, the MB isoenzyme is absent or makes only a transient appearance in the serum (Dixon *et al.*, 1973). The small rise in the activity of the MB isoenzyme may result from direct injury to the heart in cardiac operations. A more persistent elevation of this isoenzyme or an increase after some time has elapsed since operation increased the probability that infarction has occurred.

The specificity of the MB isoenzyme of creatine kinase as an indicator of myocardial damage is also an advantage when serial changes in enzyme activity are used to estimate the volume of tissue infarcted. Release of creatine kinase from non-cardiac tissues (e.g. as a result of intramuscular injections) can complicate or even invalidate such estimates based on total creatine kinase activity. However, the precision of quantitative methods of determing the MB isoenzyme is not yet great enough for the reliable prediction of eventual infarct size from measurements made during the early stages of enzyme release. Creatine kinase isoenzymes may also be of some help in the vexing problem of the diagnosis of pulmonary embolism. Creatine kinase is elevated infrequently in this condition but, when raised activities are present, the MB isoenzyme is not found, excluding myocardial damage.

Increased activities of the MB isoenzyme of creatine kinase in serum occur frequently in progressive and congenital muscular dystrophies (Goto and Katsuki, 1970; Somer *et al.*, 1973) (Fig. 7.4). The origin of the MB isoenzyme is uncertain; the skeletal muscles themselves with their abnormal isoenzyme distribution are the most probable source, and a significant correlation has been reported in the early stages of Duchenne dystrophy between muscle degeneration as measured by the urinary creatine/creatinine ratio and the contribution of the MB isoenzyme to the total serum creatine kinase activity (Takahashi *et al.*, 1977). Since cardiac muscle is involved in progressive muscular dystrophy, this is also a possible source of the MB isoenzyme. However, no correlation was found between electrocardiogram changes and the level of MB-creatine kinase. The MB isoenzyme of creatine kinase has also been detected in the serum of patients with polymyositis where it may be related to changes in the MB isoenzyme content of muscle accompanying regeneration. As with the isoenzymes of lactate dehydrogenase, studies of creatine kinase iso-enzymes in serum probably do not contribute greatly to the investigation of muscle diseases except when the absence of the MB isoenzyme directs attention to skeletal muscle rather than the heart as the source of an unexplained elevation of total serum creatine kinase activity.

The BB isoenzyme of creatine kinase has been observed in serum in the rare inherited condition of malignant hyperpyrexia. In this condition, administration of anaesthetics may precipitate a fatal hyperpyrexia accompanied by very high levels of serum creatine kinase. Elevated activities usually persist in survivors. The unusual isoenzyme pattern seems to reflect an abnormality of the skeletal muscles, though only a minority of affected individuals show clinical myopathy.

Since BB-creatine kinase is present in high concentrations in brain tissue, it would be expected that brain injury would cause the appearance of this isoenzyme in serum. This has indeed been observed in acute brain injury and following neurosurgery, the incidence of such findings being particularly high when sensitive radioimmunoassays are used (Phillips *et al.*, 1980). However, several reports have shown that in cerebrovascular accidents and psychotic disorders the serum creatine kinase may be raised due to an excess of the MM isoenzyme, presumably released from skeletal muscle. Other tissues such as smooth muscle, prostate, kidney, thyroid and lung contain the BB isoenzyme, although in smaller amounts than brain, and leakage of the isoenzyme from these tissues may account for the elevation of BB isoenzyme in serum which has been observed in renal tubular necrosis, renal failure treated by dialysis, and Reye's syndrome (a fatty degeneration of the viscera), and in patients undergoing coronary artery bypass grafting (Vladutiu *et al.*, 1977). Therefore, the appearance of BB creatine kinase in serum cannot be regarded as an infallible or specific indication of brain damage. Recent studies suggest that the presence of the BB isoenzyme of creatine kinase may also be a valuable indication of malignant disease of breast and prostate (Thompson *et al.*, 1980; Feld *et al.*, 1980; Jockers-Wretou *et al.*, 1980).

MULTIPLE FORMS OF ALKALINE PHOSPHATASE

Measurements of increases in alkaline phosphatase activity in serum have been of major diagnostic value in diseases of bone and liver for nearly 50 years. For more than half that period, the problem of the identity or dissimilarity of alkaline phosphatases from various tissues was extensively debated. Until an answer to this question was supplied by the development of isoenzyme techniques in the early 1960s it was impossible also to decide finally between the two competing theories of the origin of the raised serum alkaline phosphatase activity in hepatobiliary disease, the retention theory and the hepatogenic theory, described in the preceding chapter. Now that significant differences between alkaline phosphatases from several tissues have been established, characterization of serum alkaline phosphatase has become one of the most useful applications of isoenzyme techniques in diagnostic enzymology (Moss, 1975).

The greatest clinical interest attaches to the detection of increased levels of liver or bone alkaline phosphatases in serum and to differentiation between these two tissues as possible sources of an elevated serum alkaline phosphatase activity. However, the application of isoenzyme techniques has also shown that intestinal alkaline phosphatase can enter the circulation in health and disease, and that the presence of the placental isoenzyme accounts for the elevated serum alkaline phosphatase activity which is frequently observed in the later stages of pregnancy. An entirely unexpected discovery made with the aid of isoenzyme techniques was that placental phosphatase-like isoenzymes may occur in sera from a proportion of patients with malignant disease, while other variant alkaline phosphatases have also been detected in serum (Chapter 6).

Virtually all the analytical methods listed in Chapter 3 have been applied to the identification and measurement of multiple forms of alkaline phosphatase in serum. Placental and intestinal alkaline phosphatases can be determined quantitatively in serum by inhibition with L-phenylalanine and, in addition, the pronounced heat stability of placental phosphatase provides a convenient approach to measurement of this isoenzyme. Discrimination between liver and bone alkaline phosphatases is more difficult, however, because of the general similarity in properties between these forms and it is correspondingly difficult to arrive at quantitative estimates of their respective activities in serum samples. Zone electrophoresis on various supporting media under appropriate conditions can separate the two enzyme forms to an extent which allows their relative activities to be assessed visually, but when the patterns are scanned in a densitometer the two peaks are usually seen to overlap to a greater or lesser extent, rendering quantitation difficult (Fig. 7.5).

Selective inactivation procedures which exploit the greater susceptibility of bone alkaline phosphatase to denaturation compared with the liver enzyme can give both qualitative and quantitative information about the isoenzyme composition of serum. Heating or exposure to concentrated solutions of urea are used as inactivating agents. Semi-quantitative information as to the relative proportions of liver and bone alkaline phosphatases is obtained by measuring the proportion of the total activity which survives a single, fixed period of exposure to the denaturing agent; e.g., if less than 20% of the total activity remains after heating at 56°C for 10 min, the serum probably contains an excess of bone phosphatase over the liver enzyme. Similar conclusions can be drawn from measurements of residual alkaline phosphatase activity after treatment of the serum sample with a 6M solution of urea under defined conditions, since the inactivating effect of urea is more pronounced when a preponderance of the bone enzyme is present.

If the progress of inactivation of alkaline phosphatase in serum is expressed as a function of time, e.g. by measuring residual activity after increasing periods of incubation at 56°C, the resulting curve can be analysed into components which correspond to the decay of the

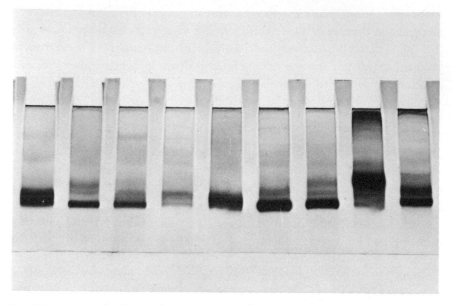

Fig. 7.5   *Polyacrylamide gel electrophoresis of alkaline phosphatase in human serum. The compact, rapidly anodally-migrating zone of liver phosphatase is visible in all samples and is especially prominent in the sixth sample from the left. The rather slower, more diffused zone of bone phosphatase predominates in samples 1 and 5. The activities of both zones are increased in sample 9. Sample 8 contains a particularly prominent intestinal phosphatase zone. The anode is at the bottom.*

respective phosphatase forms and their initial activities can be calculated (Whitby and Moss, 1975). The half-lives of liver and bone alkaline phosphatases at 56°C are not constant from serum to serum, although the half-life of liver phosphatase is always about four times as long as that of the bone enzyme. This variation in half-lives probably results at least partly from differences between sera in factors such as pH which affect the stability of the enzymes (Moss *et al.*, 1972). Its existence introduces errors into methods in which a constant value of half-life is used to calculate the activity of liver phosphatase from the residual activity measured after a single period of inactivation by heat.

Determination of the full progress curve of inactivation by heat for each serum sample eliminates errors due to sample-dependent variations in enzyme stability but is too laborious and time-consuming for routine clinical analysis, although an automated method of generating such curves has been described (PetitClerc, 1976). A satisfactory compromise between the determination of multipoint inactivation curves and single-period inactivation procedures which largely avoids the sample-dependent errors of the latter approach is to measure residual activities after two different periods of inactivation by heat (Moss and Whitby, 1975). The two periods are selected so that the activity measured after each incubation is due almost entirely to liver alkaline phosphatase. This method has been evaluated extensively in diagnostic practice (Whitaker *et al.*, 1977; 1978). The results confirm that, as expected, greater sensitivity is obtained by the measurement of individual enzyme forms than by determination of total enzyme activity. The increase in sensitivity is of the order of two-fold in both bone and liver diseases (Fig. 7.6).

Slight or moderate elevations in serum alkaline phosphatase activity are not infrequently detected during routine biochemical investigation and may be the only apparent abnormality in the composition of the blood: the need to differentiate between liver and bone as alternative possible

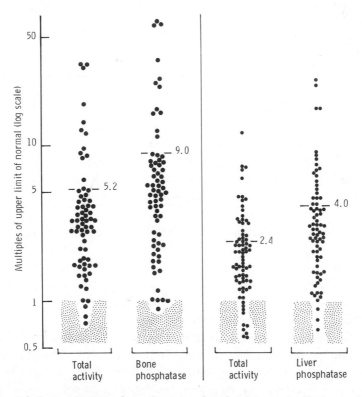

Fig. 7.6   *Relative sensitivities of measurements of total and tissue-specific forms of alkaline phosphatase in serum in hepatobiliary disease (right) and Paget's disease of bone (left) determined by a quantitative heat-inactivation technique (Moss and Whitby, 1975). Because of the marked dependence of activity on age, elevations are expressed as multiples of the appropriate upper reference limit in each case.*

sources of the increased enzyme activity in such cases accounted for nearly half of a series of requests for alkaline phosphatase characterization (Whitaker *et al.*, 1978). Problems of this nature can generally be solved by qualitative techniques, such as the inspection of electrophoretic patterns. However, quantitative methods are preferable, or even essential, when the elevation of total alkaline phosphatase activity is minimal. Furthermore, an abnormal level of either liver or bone phosphatase may be found by quantitative analysis of sera from patients in whom disease of liver or bone is suspected, but whose total serum alkaline phosphatase activity is within normal limits. In a series of nearly 500 samples from adults and children, about 30 were found to contain increased activities of either the bone or liver enzyme although the total alkaline phosphatase activities were not elevated (Whitaker *et al.*, 1978). Quantitative methods also allow changes in the activities of different phosphatases to be monitored during treatment.

The ability to measure absolute activities of liver or bone phosphatases is particularly valuable when the activity of one phosphatase is already raised, whether for physiological reasons as in normal growth or pregnancy, or because of established disease in either bone or liver, since small changes in the level of the second phosphatase are not detectable with certainty by qualitative methods. Problems of this nature are frequently encountered in the investigation and treatment of malignant disease, in deciding whether liver or bone, or both these tissues, have been infiltrated by the disease, and their solution constitutes the greatest single contribution of alkaline phosphatase characterization to medicine (Fig. 7.7).

The ectopic production of apparently normal or modified forms of alkaline phosphatase by tumours has already been discussed in the preceding chapter. In view of the low incidence of these isoenzymes in the sera of patients with cancers of various types (estimated at between 3–15% depending on the sensitivity of the methods used), a diagnosis of malignant disease will only rarely be based on the demonstration of a tumour-derived alkaline phosphatase in serum. Nevertheless, the identification and quantitative measurement of these isoenzymes are valuable in interpreting raised serum alkaline phosphatase levels in cancer and in following the effects of treatment in those patients in whom these isoenzymes are present. Isoenzymes of alkaline phosphatase resembling the normal placental isoenzyme have been detected in sera from patients with a wide variety of different cancers. Apart from the association of a fast variant with hepatoma, there seems to be little correlation between

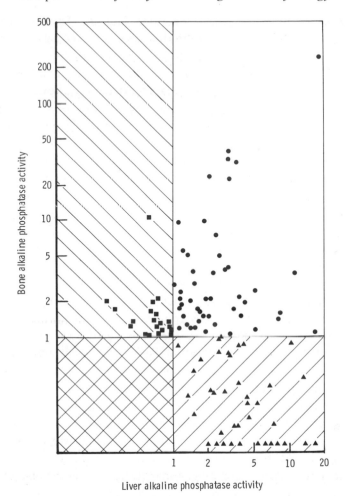

Fig. 7.7 *Elevations of bone and liver alkaline phosphatases in sera from patients with malignant diseases. Quantitative measurement of the individual enzyme forms by a selective heat-inactivation technique (Moss and Whitby, 1975) allows patients in whom only the bone (■) or liver enzyme (▲) is elevated, or in whom both phosphatases are raised (●) to be identified. Values are expressed as multiples of the appropriate upper reference limits, corrected for age.*

the type of cancer and the appearance of the isoenzymes in serum. However, they are rather more common in cancers of the ovary or testis compared with those of lung or breast.

Carcinoplacental isoenzymes of alkaline phosphatase can display a range of electrophoretic mobilities, as do the allelic variants of normal placental alkaline phosphatase: electrophoresis alone is therefore an unreliable method for their detection. The stabilities of the Regan isoenzyme and the L-leucine inhibitable Nagao form equal or approach the pronounced stability to heat of normal placental alkaline phosphatase, so that the measurement of residual alkaline phosphatase activity after serum has been heated at 65°C is a more reliable procedure. The carcinoplacental isoenzymes also share the antigenic determinants of normal placental alkaline phosphatase and react with antiserum to placental phosphatase. Consequently, immunochemical methods can be used to identify Regan and similar isoenzymes in serum. Careful characterization by several techniques is necessary to distinguish isoenzymes of a non-placental type which may originate in tumours from their normal counterparts in liver, bone or kidney.

The presence of normal placental alkaline phosphatase accounts for the elevated serum alkaline phosphatase activity which is a common finding during the last third of normal pregnancy. Quantitative measurement of the contribution of the placental isoenzyme to the total activity in serum, e.g. by heat stability measurements, L-phenylalanine inhibition or immunochemical means, allows concomitant changes in the activity of other isoenzymes to be detected and interpreted in the usual way. The activity in serum of the placental isoenzyme itself reflects placental function, and changes in placental phosphatase activity can help in the detection of such conditions as hypertension and pre-eclampsia.

However, the range of placental alkaline phosphatase activity in the serum of healthy pregnant women is very wide, so that the interpretation of isolated measurements of this isoenzyme during pregnancy is difficult or impossible. More reliable information can be derived from the trend of sequential determinations; e.g. a sharply upward trend may presage the onset of pre-eclampsia or threatened abortion.

A zone of alkaline phosphatase corresponding in electrophoretic mobility to the isoenzyme of small intestine can be recognized in about a third of serum samples from healthy subjects. The presence of the zone of intestinal alkaline phosphatase in serum is correlated with ABO blood group and secretor status, and its intensity with diet (Langman *et al.*, 1966). Secretors of blood group substances who are of blood groups B or O are more likely to have an intestinal phosphatase zone in the serum than individuals of other groups or non-secretors. The presence of the intestinal isoenzyme increases the average normal serum alkaline phosphatase activity somewhat compared with the average level in normal subjects without this isoenzyme in their sera.

Increased activities of intestinal alkaline phosphatase in serum have been noted both in intestinal diseases and in non-intestinal disea; as cirrhosis of the liver (Fishman *et al.*, 1965). Elevated levels of ii alkaline phosphatase have also been noted in sera from patients on chronic haemodialysis (Walker, 1974a). In some cases, the intestinal isoenzyme may account for the major part of the total alkaline phosphatase activity of the serum. However, an elevated intestinal phosphatase component in serum is not a constant finding, even in patients with apparently similar diseases. The factors which influence the entry of intestinal alkaline phosphatase into the circulation (or possibly its greater persistence in the plasma of some individuals: recent observations (Bayer *et al.*, 1980) indicate that intestinal phosphatase is less firmly bound by erythrocytes of blood groups B and O than by those of group A) in health and disease and the relationship of these factors to blood type are not understood. Therefore, measurements of the intestinal alkaline phosphatase activity of serum are of limited diagnostic value at present.

In hepatobiliary disease of all types in which serum alkaline phosphatase activity is elevated, the main zone of activity seen on electrophoresis of serum corresponds to the major zone visible in extracts of whole liver. The intensification of this zone in serum in liver disease is now attributed to new synthesis of alkaline phosphatase, chiefly in parenchymal cells (Chapter 6). Additional minor alkaline phosphatase zones which have a high net charge, but which migrate only slowly towards the anode in some gel media because of their high molecular weight, are often present in sera from patients with hepatobiliary disease. It is probable that these zones represent aggregates of molecules of liver phosphatase with other components such as lipoproteins. Aggregates of alkaline phosphatase occur in bile, and regurgitation of bile alkaline phosphatase, particularly in post-hepatic obstruction, may account for the presence of some zones of high molecular weight in serum. Also, alkaline phosphatase, as well as other enzymes, appears to enter the circulation attached to fragments of cell membranes in liver disease (Shinkai and Akedo, 1972; Borgers *et al.*, 1975). It is possible that the occurrence of these enzyme forms may have diagnostic value (Crofton *et al.*, 1979). As is the case with a number of other enzymes, alkaline phosphatase occasionally forms high molecular-weight complexes with immunoglobulins, but no specific clinical significance can be attached to the appearance of these complexes.

ACID PHOSPHATASE ISOENZYMES

The characterization of the forms of this enzyme in serum constitutes the earliest application of isoenzyme techniques in diagnostic enzymology, although it antedates by two decades the concept of tissue-specific enzyme variation within a single organism as a generalized phenomenon. The stimulus to study the properties of the enzyme in serum was provided by the need to identify reliably small increases in serum acid phosphatase activity deriving from prostatic cells, indicating metastatic spread of carcinoma of the prostate gland, against a small, but in relative terms highly significant, background of acid phosphatase activity originating in cells of other types. Thus both of the diagnostic requirements which can potentially be met by the use of isoenzyme techniques in diagnostic enzymology, namely greater sensitivity and increased tissue-specificity, were present in developing the diagnostic usefulness of determinations of serum acid phosphatase activity. Acid phosphatase measured by its catalytic activity is increased in the sera of about 75% of patients with metastatic carcinoma of the prostate, whereas levels are typically normal while the disease remains confined to the prostate gland, or in benign prostatic hypertrophy.

Differences in catalytic properties between prostatic acid phosphatase and those of other isoenzymes which occur in serum have already been mentioned (Chapter 3) and, until recently, these differences have formed the sole basis of attempts to increase the sensitivity and specificity of measurements of prostatic acid phosphatase in serum. More recently, the availability of anti-prostatic phosphatase antisera has made possible the application of a variety of immunoassays, including radioimmunoassay.

The sensitivity and specificity with which small increases in acid phosphatase activity in serum can be detected and measured is improved by the use of the specific inhibitor, L-(+)-tartrate (Fig. 7.8), or by the use of substrates such as α-naphthyl phosphate which are less readily hydrolysed by non-prostatic, tartrate-resistant isoenzymes (Fig. 7.9). However, even these refinements are generally unable to detect carcinoma *in situ*. Among various immunoassays which have been described, the double-antibody immunoprecipitation method of Choe *et al.* (1980) appears to be capable of improving the precision of measurements of low activities of prostatic acid phosphatase, while at the same time reducing the background of non-prostatic acid phosphatase activity (Fig. 7.10).

The immunoprecipitation method depends on the retained enzymic activity of the isoenzyme–antibody complex. A potential advantage of radioimmunoassay is that the isoenzyme molecules do not need to be catalytically active in order to be measured, provided that they are immunologically recognized by the specific antibodies; therefore, this

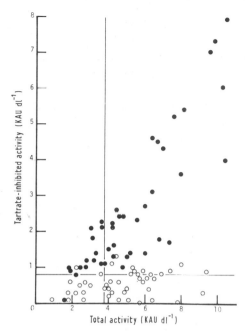

Fig. 7.8   *Discrimination between elevations of serum acid phosphatase activity due to carcinoma of the prostate (solid symbols) and similar elevations due to osteolytic bone diseases (open symbols) by inhibition of the prostatic isoenzyme with dextrorotatory tartrate. The substrate is phenyl phosphate. Horizontal and vertical lines indicate the appropriate upper reference limits (From Moss, 1977. By permission of Pergamon Press).*

method may give access to a pool of circulating isoenzyme molecules which may be considerably greater than that composed of catalytically active molecules, if partially-altered molecules are also present, so increasing the sensitivity of detection of tumour products. Although first reports of the use of radioimmunoassays suggested that the technique is able to detect a considerable proportion of cases of prostatic carcinoma in their early stages, before metastases appear, later assessments are less optimistic. This indicates perhaps that inactive acid phosphatase molecules are not present in the circulation in significant amounts in non-metastatic disease. Nevertheless, while the use of radioimmunoassay as a screening procedure for the detection of prostatic cancer in symptom-free subjects may not prove to be as effective as had been hoped, improvements in the determination of prostatic acid phosphatase made possible by immunoassays represent an important extension of the value of the isoenzyme in the investigation and management of cancer of the prostate (Watson and Tang, 1980).

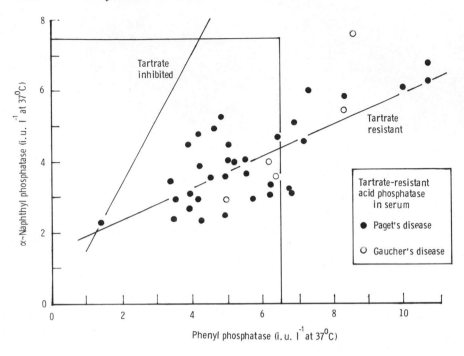

Fig. 7.9   *Comparison of acid phosphatase activities in sera from patients with Paget's disease of bone or Gaucher's disease of spleen determined with two substrates. The values are essentially normal with α-naphthyl phosphatase as substrate but several are slightly or moderately elevated with phenyl phosphate. The slope of the regression line for the tartrate-resistant isoenzymes present in these conditions is significantly different from that found for the tartrate-inhibitable prostatic isoenzyme. The vertical and horizontal lines indicate upper reference limits for the two methods.*

Slight or moderate elevations of serum acid phosphatase activity also occur in some non-prostatic diseases. It was recognized early in the clinical use of this enzyme test that such elevations were seen less often with some substrates than with others, implying a difference in properties between prostatic and non-prostatic acid phosphatases. Such elevations are often present in various bone diseases, such as Paget's disease of bone (Fig. 7.9), hyperparathyroidism with skeletal involvement, and when skeletal metastases of certain cancers are present. The number and activity of osteoclasts is increased in these conditions and, since these cells are rich in acid phosphatase, they are assumed to be the source of the increased acid phosphatase of the serum. These cells are also the probable source of the slightly higher levels of serum acid phosphatase that are observed in

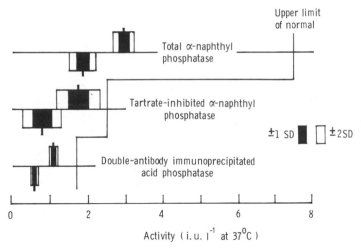

Fig. 7.10 *Upper limits of normal and sensitivities of three methods of determining prostatic acid phosphatase activity in serum. Sensitivity was estimated from the precision of measurements of activity before and after addition of a small amount of the prostatic isoenzyme to serum. Compared with measurement of total α-naphthyl phosphatase activity, the use of tartrate reduces the background acid phosphatase activity against which the increment of prostatic phosphatase must be detected but at the cost of reduced precision. The double-antibody immunoprecipitation method provides both greater specificity for the prostatic isoenzyme (i.e. reduced background activity) and improved precision.*

growing children compared with normal adult levels. The 'bone' acid phosphatase is not inhibited by tartrate. Its appearance in the circulation is of little diagnostic value, although it has been considered to be of some help in detecting skeletal metastases in cancer of the breast and in determining the effectiveness of treatment.

A tartrate-resistant form of acid phosphatase is frequently increased in sera from patients with Gaucher's disease, the most common of the sphingolipid storage diseases (Chapter 5). The raised activity is most apparent when substrates such as phenyl phosphate are used (Fig. 7.9). Lipid-laden histocytes (Gaucher cells) are present in the spleen and throughout the reticuloendothelial system in Gaucher's disease and these cells are rich in acid phosphatase. Two forms of acid phosphatase can be separated by ion-exchange chromatography of extracts of Gaucher's disease spleen, one of which resembles in several respects the major fraction of acid phosphatase present in serum in this condition (Robinson and Glew, 1980). However, a tartrate-inhibited acid phosphatase which is also present in increased amounts in spleen in Gaucher's disease is less prominent in the serum, so that some process of selective release from the tissue or elimination from the circulation may be presumed to operate (Robinson and Glew, 1980).

OTHER ISOENZYMES

The mitochondrial isoenzyme of *aspartate aminotransferase* has been detected together with the cytoplasmic form in the sera of patients with diseases of the heart or liver, whereas only the cytoplasmic isoenzyme occurs in normal serum. The occurrence of significant amounts of the mitochondrial isoenzyme in such conditions as myocardial infarction and in severe hepatitis and active cirrhosis has been interpreted as reflecting the extent of cellular necrosis in these diseases, in contrast to conditions in which inflammation predominates over necrosis and which result in a selective loss of cytoplasmic aspartate aminotransferase from the cell (Schmidt *et al.*, 1967; Boyde, 1968). Similarly, the serum aspartate aminotransferase activity in progressive muscular dystrophy is largely in the form of the cytoplasmic isoenzyme and this favours increased membrane permeability, rather than necrosis of muscle cells, as the underlying cause of the increased enzyme activities in serum in this condition (Somer *et al.*, 1973).

These observations illustrate the use of isoenzyme studies to investigate subcellular changes in pathological states. However, the day-to-day diagnostic applications of fractionation of aspartate aminotransferase isoenzymes in serum have not proved to be extensive. Pathological conditions which are accompanied by release of aspartate aminotransferase into the circulation are usually not clearly separable into exclusively necrotic or inflammatory categories, and an estimate of the relative contributions of these two processes in any one case is of little diagnostic value.

Salivary and pancreatic isoenzymes of α-*amylase* in serum have been separated by zone electrophoresis, chromatography and gel-filtration. Quantitative analysis has been carried out by densitometry of electrophoretic patterns, or by selective inhibition or inactivation, and a radioimmunoassay has been devised for salivary amylase (Boehm-Truitt *et al.*, 1978).

The main diagnostic importance of serum amylase measurements lies in the investigation of acute pancreatitis, in which hyperamylasaemia is a valuable diagnostic sign. Measurements of total serum amylase are less useful in chronic pancreatic disease of various kinds since normal levels are typically found. However, as with other applications of isoenzyme studies in diagnosis, specific quantitative measurement of pancreatic amylase may provide useful information about altered pancreatic function by revealing a significant decrease in the activity of this isoenzyme even though the total serum amylase activity is not markedly reduced (Berk and Fridhandler, 1975; Wolf *et al.*, 1976; Skude and Kollberg, 1976; Warshaw, 1977; Gillard and Feig, 1979). The salivary isoenzyme is increased in serum in diseases affecting the salivary glands, such as mumps and parotitis, but these tissues are not the only source of the isoenzyme and elevations have also been reported in some cases of ovarian cancer and chronic hepatitis (Lehrner *et al.*, 1976; Takeuchi *et al.*, 1974).

Amylase was the first enzyme for which post-translational modification by formation of enzyme–immunoglobulin complexes *in vivo* was detected; as for other examples of this phenomenon, however, few diagnostically useful conclusions can be drawn from the occurence of such macro-amylase complexes.

Considerable clinical importance attaches to the identification of tumour products which may signal the existence and location of malignant disease by their appearance in the circulation. Many such tumour markers have been identified and assessed. Prostatic acid phosphatase is outstanding among those which are enzymic, the useful-ness of carcinoplacental isoenzymes of alkaline phosphatase being limited by their low incidence in the sera of cancer patients and that of the BB isoenzyme of creatine kinase by its occurrence in non-malignant diseases. Many studies have been made of changes in the composition of cell surfaces accompanying malignant transformation and circulating markers of these changes have been sought. Glycosyl transferases of various kinds are among the membrane-derived substances which may be detectable in the blood of cancer patients, and a high incidence of a particular form of galactosyltransferase, 'isoenzyme II', has been found in sera from patients with a variety of malignant diseases (Podolsky *et al.*, 1978). Isoenzyme II from human sources resembles the normal form of galactosyltransferase in several kinetic properties, but differs in its Michaelis constant for the galactosyl receptor. It also has a significantly greater molecular weight than the normal form and is chromatographically and electrophoretically distinct (Podolsky and Weiser, 1980).

When allelic variation of enzymes results in disease, the catalytic activity of the mutant enzyme is usually very low or even zero, so that demonstration of a low or absent total enzyme activity in an appropriate tissue sample confirms the diagnosis, without the need for, or possibility of, isoenzyme characterization in serum samples. Exceptions to this generalization may arise in those conditions in which the existence of the mutant enzyme, and its attendant consequences, only become apparent on exposure to some environmental or therapeutic hazard. In these circumstances, isoenzyme characterization may allow the probable response to particular conditions to be anticipated and possible risks to be reduced. The genetically-determined variants of serum cholinesterase already described provide examples of this application of isoenzyme characterization.

References

Abul-Fadl, M.A.M. and King E.J. (1948), *J. Clin. Pathol.*, 1, 80–90.

Abul-Fadl, M.A.M. and King, E.J. (1949), *Biochem. J.*, 45, 51–60.

Adinolfi, A. and Hopkinson, D.A. (1978), *Ann. Hum. Genet.*, 41, 399–407.

Ahmed, Z. and King, E.J. (1960), *Biochim. Biophys. Acta*, 45, 581–592.

Andersson, B., Nyman, P.O. and Strid, L. (1972), *Biochem. Biophys. Res. Commun.*, 48, 670–677.

Appella, E. and Markert, C.L. (1961), *Biochem. Biophys. Res. Commun.*, 6, 171–176.

Applebury, M.L., Johnson, B.P. and Coleman, J.E. (1970), *J. Biol. Chem.*, 245, 4968–4976.

Armstrong, J.B., Lowden, J.A. and Sherwin, A.L. (1977), *J. Biol. Chem.*, 252, 3112–3116.

Augustinsson, K.-B. (1961), *Ann. NY Acad. Sci.*, 94, 844–860.

Babson, A.L., Read, P.A. and Phillips, G.E. (1959), *Amer. J. Clin. Pathol.*, 32, 88–91.

Badger, K.S. and Sussman, H.H. (1976), *Proc. Natl. Acad. Sci. USA*, 73, 2201–2205.

Baker, R.W.R. and Pellegrino, C. (1954), *Scand. J. Clin. Lab. Invest.*, 6, 94–101.

Barker, J.S.F. and East, P.D. (1980), *Nature*, 284, 166–169.

Bayer, P.M., Hotschek, H. and Knoth, E. (1980), *Clin. Chim. Acta*, 108, 81–87.

Bazzell, K.L., Price, G., Tu, S., Griffin, M., Cox, R. and Ghosh, N. (1976), *Europ. J. Biochem.*, 61, 493–499.

Beautyman, W. (1962), *Lancet*, ii, 305.

Beckman, G., Beckman, L. and Tärnvik, A. (1970), *Hum. Hered.*, 20, 81–85.

Belliveau, R.E., Wiernik, P.H. and Stickles, E.A. (1973), *Lancet*, i, 22–24.

Berg, T. and Blix, A.S. (1973), *Nature New Biol.*, 245, 239–240.

Berk, A.J. and Clayton, D.A. (1973), *J. Biol. Chem.*, 248, 2722–2729.

Berk, J.E. and Fridhandler, L. (1975), *Amer. J. Gastroenterol.*, 63, 457–463.

Bernstein, L.H. (1977), *Clin. Chem.*, 23, 1928–1930.

Beutler, E. (1978), In *The Metabolic Basis of Inherited Disease*, 4th edn, (Eds. Stanbury, J.B., Wyngaarden, J.B. and Fredrickson, D.S.), McGraw Hill, New York, pp. 1430–1451.

Beutler, E. and Kuhl, W. (1975), *Nature*, 258, 262–264.

Beutler, E., Yoshida, A., Kuhl, W. and Lee, J.E.S. (1976), *Biochem. J.*, 159, 541–543.

Biwenga, J. (1973), *Clin. Chim. Acta*, 47, 139–147.

Biwenga, J. (1977), *Clin. Chim. Acta*, 76, 149–153.

Blanchaer, M.C. (1961), *Clin. Chim. Acta*, 6, 272–275.

Blomberg, F., and Raftell, M. (1974), *Europ. J. Biochem.*, 49, 21–29.

Bodansky, O. (1937), *J. Biol. Chem.*, 118, 341–361.

Boehm-Truitt, M., Harrison, E., Wolf, R.O. and Notkins, A.L. (1978), *Analyt. Biochem.*, **85**, 476–487.

Bøg-Hansen, T.C. and Daussant, J. (1974), *Analyt. Biochem.*, **61**, 522–527.

Borgers, M., De Broe, M.E. and Wieme, R.J. (1975), *Clin. Chim. Acta*, **59**, 369–372.

Borisov, V.V., Borisova, S.N., Kachalova, G.S., Sosfenov, N.I., Vainshtein, B.K., Torshinsky, Y.M. and Braunstein, A.E. (1978), *J. Mol. Biol.*, **125**, 275–292.

Bossa, F., Polidoro, G., Barra, D., Liverzani, A. and Scandurra, R. (1976), *Internatl. J. Peptide and Protein Res.*, **8**, 499–501.

Boyd, J.W. (1961), *Biochem. J.*, **81**, 434–441.

Boyd, J.W. (1962), *Clin. Chim. Acta*, **7**, 424–431.

Boyde, T.R.C. (1968), *Enzymol. Biol. Clin.*, **9**, 385–392.

Boyer, S.H. (1963), *Ann. N.Y. Acad. Sci.*, **103**, 938–950.

Boyer, S.H., Faines, D.C. and Watson-Williams, E.J. (1963), *Science*, **141**, 642–643.

Brady, R.O. (1978), In *The Metabolic Basis of Inherited Disease*, 4th Edn., (Eds. Stanbury, J.B., Wyngaarden, J.B. and Fredrickson, D.S.), McGraw Hill, New York, pp. 731–746.

Brinster, R.L. (1979), *Isozymes: Current Topics in Biological and Medical Research*, **3**, 155–184.

Brydon, W.G., Crofton, P.M., Smith, A.F., Barr, D.G.D. and Harkness, A.R. (1975), *Biochem. Soc. Trans.*, **3**, 927–929.

Burger, A., Richterich, R. and Aebi, H. (1964), *Biochem. Z.*, **339**, 305–314.

Butterworth, P.J., and Moss, D.W. (1966), *Nature*, **209**, 805–806.

Cahn, R.D., Kaplan, N.O., Levine, L. and Zwilling, E. (1962), *Science*, **136**, 962–969.

Campbell, D.M. and Moss, D.W. (1962), *Proc. Assoc. Clin. Biochemists*, **2**, 10–11.

Cardenas, J.M. and Dyson, R.D. (1973), *J. Biol. Chem.*, **248**, 6938–6944.

Cardenas, J.M., Dyson, R.D. and Strandholm, J.J. (1975). In *Isozymes*, Vol. 1 (Ed. Markert, C.L.), Academic Press, New York, pp. 523–541.

Carleer, J. and Ansay, M. (1976), *Internatl. J. Biochem.*, **7**, 565–567.

Carlier, M.-F. and Pantaloni, D. (1973), *Europ. J. Biochem.*, **37**, 341–354.

Carney, J.A. (1976), *Clin. Chim. Acta*, **67**, 153–158.

Carroll, M., Dance, N., Masson, P.K., Robinson, D. and Winchester, B.G. (1972), *Biochem. Biophys. Res. Commun.*, **49**, 572–583.

Carter, N., Jeffery, S., Shiels, A., Edwards, Y., Tipler, T. and Hopkinson, D.A. (1979), *Biochem. Genet.*, **17**, 837–854.

Chang, C.-H., Ruam, S., Angellis, D., Doellgast, G. and Fishman, W.H. (1975), *Cancer Res.*, **35**, 1706–1712.

Chang, S.-M., Lee, C.-Y. and Li, S.S.-L. (1979), *Biochem. Genet.*, **17**, 715–729.

Chao, S.-E., and Scandalios, J.G. (1975), In *Isozymes*, Vol 3., (Ed. Markert, C.L.), Academic Press, New York, pp. 675–690.

Chen, S.H. and Giblett, E.R. (1971), *Amer. J. Hum. Genet.*, **23**, 419–424.

Chester, M.A., Lundblad, A. and Masson, P.K. (1975), *Biochim. Biophys. Acta*, **391**, 341–348.

Cho, H.W., Meltzer, H.Y., Joung, J.L. and Goode, D. (1976), *Clin. Chim. Acta*, **73**, 257–265.

Choe, B.K., Pontes, E.J., Dong, M.K. and Rose, N.R. (1980), *Clin. Chem.*, **26**, 1854–1859.

Choe, B.-K., Pontes, E.J., Rose, N.R. and Henderson, M.D. (1978), *Investigative Urology*, **15**, 312–318.

Chu., T.M., Wang, M.C., Scott, W.W., Gibbons, R.P., Johnson, D.E., Schmidt, J.D., Loening, S.A., Prout, G.R. and Murphy, G.P. (1978), *Investigative Urology*, **15**, 319–323.

Clubb, J.S., Neale, F.C. and Posen, S. (1965), *J. Lab. Clin. Med.*, **66**, 493–507.

Cohen, P. (1976), *Control of Enzyme Activity*, Outline Studies Series, Chapman and Hall, London.

Conyers, R.A.J., Birkett, D.J., Neale, F.C., Posen, S. and Brudenell-Woods, J. (1967), *Biochim. Biophys. Acta*, **139**, 363–371.

Cox, R.P., Elson, N.A., Tu, S. and Griffin, M.J. (1971), *J. Mol. Biol.*, **58**, 197–215.

Criss, W.E. (1971), *Cancer Res.*, **31**, 1523–1542.

Crofton, P.M., Elton, R.A. and Smith, A.F. (1979), *Clin. Chim. Acta*, **98**, 263–275.

Crofton, P.M. and Smith, A.F. (1978), *Clin. Chim. Acta*, **83**, 235–247.

Curthoys, N.P. and Hughey, R.P. (1979), *Enzyme*, **24**, 383–403.

Dale, G. and Latner, A.L. (1968), *Lancet*, **i**, 847–848.

Davidson, R.G. and Cortner, J.A. (1967), *Science*, **157**, 1569–1571.

Davidson, R.G., Cortner, J.A., Rattazzi, M.C. Ruddle, F.H. and Lubs, H.A. (1970), *Science*, **169**, 391–392.

Davidson, R.G., Nitowsky, H.M. and Childs, B. (1963), *Proc. Natl. Acad. Sci. USA*, **50**, 481–485.

Davies, D.R. (1934), *Biochem. J.*, **28**, 529–536.

Davis, B.J. (1964), *Ann. NY Acad. Sci.*, **121**, 404–427.

Dawson, D.M., Eppenberger, H.M. and Eppenberger, M.E. (1968), *Ann. NY Acad. Sci.*, **151**, 616–626.

Dawson, D.M., Eppenberger, H.M. and Kaplan, N.O. (1965), *Biochem. Biophys. Res. Commun.*, **21**, 346–353.

Dawson, D.M., Eppenberger, H.M. and Kaplan, N.O. (1967), *J. Biol. Chem.*, **242**, 210–217.

Dawson, D.M. and Greene, J.M. (1975), In *Isozymes* Vol. 1, (Ed. Markert, C.L.). Academic Press, New York, pp. 381–388.

Delbruck, A. and Henkel, E. (1979), *Europ. J. Biochem.*, **99**, 65–69.

Delorenzo, R.J. and Ruddle, F.H. (1970), *Biochem. Genet.*, **4**, 259–273.

Dern, R.J., McCurdy, P.R. and Yoshida, A. (1969), *J. Lab. Clin. Med.*, **43**, 303–309.

Detter, J.C., Ways, P.O., Giblett, E.R., Baughan, M.A., Hopkinson, D.A., Povey, S. and Harris, H. (1968), *Ann. Hum. Genet.*, **31**, 329–338.

Deul, D.H. and Van Breemen, J.F.L. (1964), *Clin. Chim. Acta*, **10**, 276–283.

Dewald, B. and Touster, O. (1973), *J. Biol. Chem.*, **248**, 7223–7233.

Dixon, M. (1971), In *The Chemistry of Life*, (Ed. Needham, J.), Cambridge University Press, Cambridge, p. 18.

Dixon, M. and Webb, E.C. (1958), *Enzymes*, Longmans, London, p. 653.

Dixon, S.H., Limbird, L.E., Roe, C.R., Wagner, G.S., Oldham, H.N. and Sabiston, D.C. (1973), *Circulation*, **47, 48**, Suppl. 3, 137–140.

Dizik, M. and Elliott, R.W. (1977), *Biochem. Genet.*, **15**, 31–46.

Doellgast, G.J. and Benirschke, K. (1979), *Nature*, **280**, 601–602.

Doellgast, G.J. and Fishman, W.H. (1976), *Nature*, **259**, 49–51.

Doellgast, G.J. and Fishman, W.H. (1977), *Clin. Chim. Acta*, **75**, 449–454.

Donald, L.J. and Robson, E.B. (1974), *Ann. Hum. Genet.*, **37**, 303–313.

Donlon, J. and Kaufman, S. (1980), *J. Biol. Chem.*, **255**, 2146–2152.

Doonan, S., Hughes, G.J., Barra, D., Bossa, F., Martini, F. and Petruzzelli, R. (1974). *FEBS Letters*, **49**, 25–28.

Dreyfus, J.C., Demos, J., Schapira, F. and Schapira, G. (1962), *C.R. Acad. Sci.*, **254**, 4384–4386.

Eaton, R.H. and Moss, D.W. (1967), *Biochem. J.*, **105**, 1307–1312.

Echetebu, Z.O. and Moss, D.W. (1982), *Enzyme*, in press.

Echetebu, Z.O. and Moss, D.W. (1982a), *Enzyme*, in press.

Edwards, Y.H. and Hopkinson, D.A. (1977), *Isozymes: Current Topics in Biological and Medical Research*, **1**, 19–78.

Edwards, Y.H. and Hopkinson, D.A. (1979), *Ann. Hum. Genet.*, **42**, 303–313.

Edwards, Y.H. and Hopkinson, D.A. (1979a), *Ann. Hum. Genet.*, **43**, 103–108.

Edwards, Y.H., Hopkinson, D.A. and Harris, H. (1971), *Ann. Hum. Genet.*, **34**, 395–408.

Edwards, Y.H., Hopkinson, D.A., and Harris, H. (1978), *Nature*, **271**, 84–87.

Eichele, G., Ford, G.C., Glor, M. and Jamsonius, J.N. (1979), *J. Mol. Biol.*, **133**, 161–180.

Elliott, B.A. and Wilkinson, J.H. (1961), *Lancet*, **i**, 698–699.

Ellis, G. and Goldberg, D.M. (1970), *J. Lab. Clin. Med.*, **76**, 507–517.

Elson, N.A. and Cox, R.P. (1969), *Biochem. Genet.*, **3**, 549–561.

Emerson, P.M., Withycombe, W.A. and Wilkinson, J.H. (1967), *Brit. J. Haematol.*, **13**, 656–664.

Engström, L. (1964), *Biochim. Biophys. Acta*, **92**, 71–78.

Eppenberger, H.M., Dawson, D.M. and Kaplan, N.O. (1967), *J. Biol. Chem.*, **242**, 204–209.

Eppenberger, H.M., Walliman, T., Kuhn, H.J. and Turner, D.C. (1975), In *Isozymes*, Vol. 2., (Ed. Markert, C.L.), Academic Press, New York, pp. 410–420.

Estborn, N. (1959), *Nature*, **184**, 1636–1637.

Eventoff, W., Hackert, M.L., Steindel, S.J. and Rossmann, M.G. (1975), In *Isozymes*, Vol. 1, (Ed. Markert, C.L.), Academic Press, New York, pp 137–150.

Eventoff, W., Rossmann, M.G., Taylor, S.S., Torff, H.-J., Meyer, H., Keil, W. and Kiltz, H.-H. (1977), *Proc. Natl. Acad. Sci. USA*, **74**, 2677–2681.

Feld, R.D., Van Steirteghem, A.C., Zweig, M.H., Weimar, G.W., Narayana, A.S. and Whittle, D.L. (1980), *Clin. Chim. Acta*, **100**, 267–273.

Ferguson-Smith, M.A., Newman, B., Ellis, P.M., Thomson, D.M.G. and Riley, I.D. (1973), *Nature New Biol.*, **243**, 271–273.

Fine, I.H., Kaplan, N.O. and White, S. (1962), *Federation Proc.*, **21**, 409.

Fishman, W.H., Greene, S. and Inglis, N.I. (1962), *Biochim. Biophys. Acta*, **62**, 363–375.

Fishman, W.H., Inglis, N.I. and Krant, M.J. (1965), *Clin. Chim. Acta*, **12**, 298–303.

Fishman, W.H., Inglis, N.R., Greene, S., Anstiss, C.L., Gosh, N.K., Reif, A.E., Rustigian, R., Krant, M.J. and Stolbach, L.L. (1968), *Nature*, **219**, 697–699.

Fitzsimmons, J.A.W. and Doherty, M.D. (1970), *Comp. Biochem. Physiol.*, **36**, 1–20.

Fleisher, G.A., Potter, C.S., Wakim, K.G., Pankow, M. and Osborne, D. (1960), *Proc. Soc. Exp. Biol. Med.*, **103**, 229–231.

Forman, D.T., Moss, D.W. and Whitaker, K.B. (1976), *Clin. Chim. Acta*, **68**, 287–290.

Foti, A.G., Cooper J.F. and Herschman, H. (1978), *Clin. Chem.*, **24**, 140–142.

Foti, A.G., Herschman, H. and Cooper, J.F. (1975), *Cancer Res.*, **35**, 2446–2452.

Fridhandler, L., Berk, J.E. and Montgomery, K. (1974), *Clin. Chem.*, **20**, 26–29.

Fridhandler, L., Berk, J.E. and Wong, D. (1974a), *Clin. Chem.*, **20**, 22–25.

Fridovich, I. (1975), *Ann. Rev. Biochem.*, **44**, 147–159.

Fritz, P.J. and Pruitt, K.M. (1977), *Isozymes: Current Topics in Biological and Medical Research*, **1**, 125–157.

Fritz, P.J., Vesell, E.S., White, E.L. and Pruitt, K.M. (1969). *Proc. Natl. Acad. Sci., USA*, **62**, 558–565.

Fruton, J.S. (1979), *Ann. N.Y. Acad. Sci.*, **325**, 1–15.

Fujimoto, Y., Nazarian, I. and Wilkinson, J.H. (1968), *Enzymol. Biol. Clin.*, **9**, 124–136.

Funakoshi, S. and Deutsch, H.F. (1970), *J. Biol. Chem.*, **245**, 4913–4919.

Gazith, J., Schultze, I.T., Gooding, R.H., Womack, F.C. and Colowick, S.P. (1968), *Ann. NY Acad. Sci.*, **151**, 307–331.

Geiger, B., Navon, R., Ben-Yoseph, Y. and Arnon, R. (1975), *Europ. J. Biochem.*, **56**, 311–318.

Geokas, M.C., Wollesen, F., Rinderknecht, H., and Marinson, F. (1974), *J. Lab. Clin. Med.*, **84**, 574–583.

Gerhardt, W., Ljungdahl, L., Börjesson, J., Hofvendahl, S. and Hedenäs, B. (1977), *Clin. Chim. Acta*, **78**, 29–41.

Ghosh, N.K. and Fishman, W.H. (1968), *Biochem. J.*, **108**, 779–792.

Gillard, B.K. and Feig, S.A. (1979), *Pediatric Res.*, **13**, 399.

Giraud, N., Marriq, C. and Laurent-Tabusse, G. (1974), *Biochimie*, **56**, 1031–1043.

Glew, R.W., Peters, S.P. and Christopher, A.R. (1976), *Biochim. Biophys. Acta*, **42**, 179–199.

Goedde, H.W. and Altland, K. (1968), *Ann. NY Acad. Sci.*, **151**, 540–544.

Goldberg, E. (1977), *Isozymes: Current Topics in Biological and Medical Research*, **1**, 79–124.

Goldman, R.D., Kaplan, N.O. and Hall, T.C. (1964), *Cancer Res.*, **24**, 389–399.

Goldstein, D.J. and Harris, H. (1979), *Nature*, **280**, 602–605.

Goldstein, D.J., Rogers, C.E. and Harris, H. (1980), *Proc. Natl. Acad. Sci. USA*, **77**, 2857–2860.

Goto, I. and Katsuki, S. (1970), *Clin. Chim. Acta*, **30**, 795–799.

Greene, P.J. and Sussman, H.H. (1973), *Proc. Natl. Acad. Sci. USA*, **70**, 2936–2940.

Grimm, F.C. and Doherty, D.G. (1961), *J. Biol. Chem.*, **236**, 1980–1985.

Grossberg, A.L., Harris, E.H. and Schlamowitz, M. (1961), *Arch. Biochem. Biophys.*, **93**, 267–277.

Hagerstrand, I. and Skude, G. (1976), *Scand. J. Clin. Lab. Invest.*, **36**, 127–129.

Haije, W.G. and De Jong, M. (1963), *Clin. Chim. Acta*, **8**, 620–623.

Hall, E.R. and Cottam, G.L. (1978), *Internatl. J. Biochem.*, **9**, 785–793.

Hall, N., Addis, P. and DeLuca, M. (1979), *Biochemistry*, **18**, 1745–1751.

Hammond, K.D. and Balinsky, D. (1978), *Cancer Res.*, **38**, 1323–1328.

Harris, H. (1975), *Principles of Human Biochemical Genetics*, 2nd Edn, North Holland, Amsterdam.

Harris, H. and Hopkinson, D.A. (1972), *Ann. Hum. Genet.*, **36**, 9–20.

Harris, H., and Hopkinson, D.A. (1976), *Handbook of Enzyme Electrophoresis in Human Genetics*, North Holland, Amsterdam.

Harris, H., Hopkinson, D.A. and Edwards, Y.H. (1977), *Proc. Natl. Acad. Sci. USA*, **74**, 698–701.

Harris, H., Hopkinson, D.A. and Robson, E.B. (1962), *Nature*, **196**, 1296–1298.

Harris, H. and Whittaker, M. (1961), *Nature*, **191**, 496–498.

Hayes, M.B. and Wellner, D. (1969), *J. Biol. Chem.*, **244**, 6636–6644.

Headings, V.E. and Tashian, R.E. (1970), *Biochem. Genet.*, **4**, 285–295.

Henderson, L.E., Henriksson, D. and Hyman, P.O. (1973), *Biochem. Biophys. Res. Commun.*, **52**, 1388–1394.

Henry, P.D., Roberts, R. and Sobel, B.E. (1975), *Clin. Chem.*, **21**, 844–849.

Herbert, F.K. (1944), *Biochem. J.*, **38**, xxiii–xxiv.

Hers, H.G. and Joassin, G. (1961), *Enzymol. Biol. Clin.*, **1**, 4–14.

Hess, B. and Walter, S.I. (1960), *Klin. Wochenschr.*, **38**, 1080–1088.

Higashino, K., Takahashi, Y. and Kang, K.Y. (1972), *Clin. Chim. Acta*, **40**, 67–81.

Hirschhorn, R., Levytska, V. and Parkman, R. (1974), *J. Clin. Invest.*, **53**, 33a.

Hjerten, S., Jerstedt, S. and Tiselius, A. (1965), *Analyt. Biochem.*, **11**, 219–223.

Hodson, A.W. and Latner, A.L. (1971), *Analyt. Biochem.*, **41**, 522–532.

Hodson, A.W., Latner, A.L. and Raine, L. (1962), *Clin. Chim. Acta*, **7**, 255–261.

Holliday, R., Porterfield, J.S. and Gibbs, D.D. (1974), *Nature*, **248**, 762–763.

Holliday, R. and Tarrant, G.M. (1972), *Nature*, **238**, 26–30.

Holmes, R.S. (1972), *FEBS Letters*, **28**, 51–55.

Holmes, R.S., Chew, G.K., Cooper, D.W. and VandeBerg, J.L. (1974), *Biochem. Genet.*, **11**, 25–32.

Holmes, R.S. and Masters, C.J. (1979), *Isozymes: Current Topics in Biological and Medical Research*, **3**, 53–114.

Holmgren, P.A., Stigbrand, T., Dumber, M.-G. and von Schoultz, B. (1978), *Clin. Chim. Acta*, **83**, 205–210.

Hooton, B.T. and Watts, D.C. (1966), *Biochem. J.*, **100**, 637–646.

Hopkinson, D.A. (1975), In *Isozymes*, Vol 1. (Ed. Markert, C.L.), Academic Press, New York, pp. 489–508.

Hopkinson, D.A. Coppock, J.S., Mühlemann, M.F. and Edwards, Y.H. (1974), *Ann. Hum. Genet.*, **38**, 155–162.

Hopkinson, D.A., Edwards, Y.H. and Harris, H. (1976), *Ann. Hum. Genet.*, **39**, 383–411.

Hopkinson, D.A. and Harris, H. (1969), *Ann. Hum. Genet.*, **33**, 81–87.

Hopkinson, D.A., Mestrimer, M.A., Cortner, J. and Harris, H. (1973), *Ann. Hum. Genet.*, **37**, 119–137.

Hopkinson, D.A., Spencer, N. and Harris, H. (1963), *Nature*, **199**, 969–971.

Horecker, B.L. (1975), In *Isozymes*, Vol. 1. (Ed. Markert, C.L.), Academic Press, New York, pp. 11–38.

Hunter, R.L. and Burstone, M.S. (1960), *J. Histochem. Cytochem.*, **8**, 58–62.

Hunter, R.L. and Markert, C.L. (1957), *Science*, **125**, 1294–1295.

Ibsen, K.H. (1977), *Cancer Res.*, **37**, 341–353.

Iino, S., Abe, K., Oda, T., Suzuki, H. and Sugiura, M. (1972), *Clin. Chim. Acta*, **42**, 161–165.

Islam, M., Bell, J.L. and Baron, D.N. (1972), *Biochem. J.*, **129**, 1003–1011.

Itakura, T., Watanabe, K., Shiokawa, H. and Kubo, S. (1978), *Europ. J. Biochem.*, **82**, 431–437.

IUPAC-IUB Commission on Biochemical Nomenclature (1977), *J. Biol. Chem.*, **252**, 5939–5941.

Jacobs, H., Heldt, H.W. and Klingenberg, M. (1964), *Biochem. Biophys. Research Commun.*, **16**, 516–521.

Jacobson, K.B. (1968), *Science*, **159**, 324–325.

Jacobus, W.E. and Lehninger, A.L. (1973), *J. Biol. Chem.*, **248**, 4803–4810.

Jacoby, B. and Bagshawe, K.D. (1971), *Clin. Chim. Acta*, **35**, 473–481.

Jacoby, B. and Bagshawe, K.D. (1972), *Cancer Res.*, **32**, 2413–2420.

Jermyn, M.A. and Thomas, R. (1954), *Biochem. J.*, **56**, 631–639.

Jockers-Wretou, E., Gericke, K., Pauly, H.E. and Pfleiderer, G. (1980), *Fresenius Z. Anal. Chem.*, **301**, 154–155.

Jockers-Wretou, E. and Pfleiderer, G. (1975), *Clin. Chim. Acta*, **58**, 223–232.

Jockers-Wretou, E. and Plessing, E. (1979), *J. Clin. Chem., Clin. Biochem.*, **17**, 731–737.

Johnson, G.B. (1977), *Isozymes: Current Topics in Biological and Medical Research*, 2, 1–19.

Jovin, T., Chrambach, A. and Naughton, M.A. (1964), *Analyt. Biochem.*, 9, 351–369.

Kagamiyama, H., Sakakibara, R., Wada, H., Tanase, S. and Morino, Y. (1977), *J. Biochem.*, 82, 291–294.

Kahn, A., Cottreau, D., Bernard, J.F. and Boivin, P. (1975), *Biomedicine*, 22, 539–549.

Kahn, A., Marie, J., Garreau, H. and Sprengers, E.D. (1978), *Biochim. Biophys. Acta*, 523, 59–74.

Kalow, W. and Genest, K. (1957), *Canad. J. Biochem. Physiol.*, 35, 339–346.

Kaminski, M. (1966), *Nature*, 209, 723–725.

Kanfer, J.N., Raghaven, S.S. and Mumford, R.A. (1975), *Biochim. Biophys. Acta*, 391, 129–140.

Kaplan, M.M. and Righetti, A. (1970), *J. Clin. Invest.*, 49, 508–516.

Kaplan, N.O. and Ciotti, M.M. (1961), *Ann. NY Acad. Sci.*, 94, 701–722.

Kaplan, N.O., Everse, J. and Admiraal, J. (1968), *Ann. NY Acad. Sci.*, 151, 400–412.

Karn, R.C., Rosenblum, B.B., Ward, J.C., Merritt, A.D. and Shulkin, J.D. (1974), *Biochem. Genet.*, 12, 485–499.

Katzen, H.M. and Soderman, D.D. (1975), In *Isozymes*, Vol 2, (Ed. Markert, C.L.), Academic Press, New York, pp. 797–817.

Keller, P.J., Kauffman, D.L., Allan, B.J. and Williams, B.L. (1971), *Biochemistry*, 10, 4867–4874.

Kelley, W.N., Rosenbloom, F.M., Henderson, J.F. and Seegmiller, J.E. (1967), *Proc. Natl. Acad. Sci. USA*, 57, 1735–1739.

Kiltz, H.-H., Keil, W., Griesbach, M., Petry, K. and Meyer, H (1977), *Hoppe-Seyler's Z. Physiol. Chem.*, 358, 123–127.

Kitto, G.B., Stolzenbach, F.E. and Kaplan, N.O. (1970), *Biochem. Biophys. Res. Commun.*, 38, 31–39.

Kitto, G.B., Wassarman, P.M. and Kaplan, N.O. (1966), *Proc. Natl. Acad. Sci. USA*, 56, 578–584.

Kitto, G.B. and Wilson, A.C. (1966), *Science*, 153, 1408–1410.

Klotz, I.M., Langeman, N.R. and Darnall, D.W. (1970), *Ann. Rev. Biochem.*, 39, 25–62.

Koehn, R.K. and Eanes, W.F. (1979), *Isozymes: Current Topics in Biological and Medical Research*, 3, 185–211.

Kohn, J. (1957), *Clin. Chim. Acta*, 2, 297–303.

Kolk, A.H.J., Van Kuyk, L. and Boettcher, B. (1978), *Biochem J.*, 173, 767–771.

Konttinen, A. and Halonen, P.I. (1962), *Amer. J. Cardiol.*, 10, 525–531.

Konttinen, A. and Somer, H. (1973), *Brit. Med. J.*, i, 386–389.

Kudirka, P.J., Busby, M.G., Carey, R.N. and Toren, E.C. (1975), *Clin. Chem.*, 21, 450–452.

Kun, E. and Volfin, P. (1966), *Biochem. Biophys. Res. Commun.*, 22, 187–193.

Kutscher, W. and Wörner, A. (1936), *Hoppe-Seyler's Z. Physiol. Chem.*, 236, 237–246.

La Du, B.N. and Choi, Y.S. (1975), In *Isozymes*, Vol. 2, (Ed. Markert, C.L.), Academic Press, New York, pp. 877–886.

Lai, C.Y., Chen, C. and Horecker, B.L. (1970), *Biochem. Biophys. Res. Commun.*, 40, 461–468.

Lam, K.W., Li, O., Li, C.Y. and Yam, L.T. (1973), *Clin. Chem.*, 19, 483–487.

Lam, W.K.W., Eastlund, D.T., Li, C.-Y. and Yam, L.T. (1978), *Clin. Chem.*, 24, 1105–1108.

La Motta, R.V., Woronick, C.L. and Rainfrank, R.F. (1970), *Arch. Biochem. Biophys.*, 136, 448–451.

Landau, W. and Schlamowitz, M. (1961), *Arch. Biochem. Biophys.*, **95**, 474–482.

Langman, M.J.S., Leuthold, E., Robson, E.B., Harris, J., Luffman, J.E. and Harris, H. (1966), *Nature*, **212**, 41–43.

Langvad, E. (1968). *Internatl., J. Cancer*, **3**, 17–29.

Latner, A.L. and Skillen, A.W. (1964), *J. Embryol. Exp. Morphol.*, **12**, 501–509.

Latner, A.L., Turner, D.M. and Way, S.A. (1966), *Lancet*, ii, 814–816.

Laurell, C.-B. (1965), *Analyt. Biochem.*, **10**, 358–361.

Lebherz, H.G. and Rutter, W.J. (1969), *Biochemistry*, **9**, 109–114.

Lee, C., Wang, M.C., Murphy, G.P. and Chu, T.M. (1978), *Cancer Res.*, **38**, 2871–2878.

Lee, J.E.S. and Yoshida, A. (1976), *Biochem. J.*, **159**, 535–539.

Lehmann, F.G. (1975), *Clin. Chim. Acta*, **65**, 257–261.

Lehmann, F.G. (1975a), *Clin. Chim. Acta*, **65**, 271–282.

Lehrner, L.M., Ward, J.C., Karn, R.C., Ehrlich, C.E. and Merritt, A.D. (1976), *Amer. J. Clin. Pathol.*, **66**, 576–587.

Lewontin, R.C. (1974), *The Genetic Basis of Evolutionary Change*, Columbia University Press, New York.

Li, J.J. (1972), *Arch. Biochem., Biophys.*, **150**, 812–814.

Lin, K.D. and Deutsch, H.F. (1972), *J. Biol. Chem.*, **247**, 3761–3766.

Lindsey, G.G. and Diamond, E.M. (1978), *Biochim. Biophys. Acta*, **524**, 78–84.

Loomis, W.F. (1975), In *Isozymes*, Vol. 3, (Ed. Markert, C.L.), Academic Press, New York, pp. 177–189.

Lowenstein, J.M. and Smith, S.R. (1962), *Biochim. Biophys. Acta*, **56**, 385–387.

Lucher-Wasyl, E. and Ostrowski, W. (1974), *Biochim. Biophys. Acta*, **365**, 349–359.

Lusis, A.J. and Paigen, K. (1977), *Isozymes: Current Topics in Biological and Medical Research*, **2**, 63–106.

Lutstorf, U.M. and von Wartburg, J.P. (1969), *FEBS Letters*, **5**, 202–206.

Lyon, M.F. (1961), *Nature*, **190**, 372–373.

McKenna, M.J., Hamilton, T.A. and Sussman, H.H. (1979), *Biochem. J.*, **181**, 67–73.

McKinley-McKee, J.S. and Moss, D.W. (1965), *Biochem. J.*, **96**, 583–587.

McKusick, V.A., Neufeld, E.F. and Kelly, T.E. (1978), In *The Metabolic Basis of Inherited Disease*, 4th edn., (Eds. Stanbury, J.B., Wyngaarden, J.B. and Frederickson, D.S.), McGraw-Hill, New York, pp. 1282–1307.

McKusick, V.A. and Ruddle, F.H. (1977), *Science*, **196**, 390–405.

Margolis, J. and Kenrick, K.G. (1968), *Analyt. Biochem.*, **25**, 347–362.

Marie, J., Simon, M.-P., Dreyfus, J.-C. and Kahn, A. (1981), *Nature*, **292**, 70–72.

Markel, S.F. and Janich, S.L. (1974), *Amer. J. Clin. Pathol.*, **61**, 328–332.

Markert, C.L. (1963), *Science*, **140**, 1329–1330.

Markert, C.L. and Appella, E. (1963), *Ann. NY Acad. Sci.*, **103**, 915–928.

Markert, C.L. and Møller, F. (1959), *Proc. Natl. Acad. Sci. USA*, **45**, 753–763.

Markert, C.L. and Ursprung, H. (1962), *Developmental Biol.*, **5**, 363–381.

Massarat, S. and Lang, N. (1965), *Klin. Wochenschr.*, **43**, 602–606.

Massaro, E.J. and Markert, C.L. (1969), *J. Exp. Zool.*, **168**, 223–228.

Massoulié, J. (1980), *Trends in Biochemical Sciences*, **5**, 160–164.

Masters, C.J. and Holmes, R.S. (1974), *Adv. Comp. Physiol. Biochem.*, **5**, 109–195.

Meisler, M.H. (1975), *Biochim. Biophys. Acta*, **410**, 347–353.

Meisler, M. and Rattazzi, M.C. (1974), *Amer. J. Hum. Genet.*, **26**, 683–691.

Mercer, D.W. (1974), *Clin. Chem.*, **20**, 36–40.

Michuda, C.M. and Martinez-Carrion, M. (1969), *Biochemistry*, **8**, 1095–1105.

Milisauskas, V. and Rose, N.R. (1972), *Clin. Chem.*, **18**, 1529–1531.

Milisauskas, V. and Rose, N.R. (1973), *Exp. Cell Res.*, **81**, 279–284.

Milstein, C. (1964), *Biochem. J.*, **92**, 410–421.

Morell, A.G., Gregoriadis, G. and Scheinberg, I.H. (1971), *J. Biol. Chem.*, **246**, 1461–1467.

Morin, L.G. (1976), *Clin. Chem.*, **22**, 92–97.

Moss, D.W. (1962), *Nature*, **193**, 981–982.

Moss, D.W. (1970), *FEBS Symposium*, **18**, 227–239.

Moss, D.W. (1970a), *Enzymologia*, **39**, 319–330.

Moss, D.W. (1973), *Clin. Chim. Acta*, **43**, 447–449.

Moss, D.W. (1975), *Enzyme*, **20**, 20–34.

Moss, D.W. (1977), *Molecular Aspects of Medicine*, **1**, 478–581.

Moss, D.W. (1979), *Isoenzyme Analysis*, The Chemical Society, London.

Moss, D.W. (1982), *Advances in Clinical Enzymology*, **2**, in press.

Moss, D.W., Campbell, D.M., Anagnostou-Kakaras, E. and King, E.J. (1961), *Biochem. J.*, **81**, 441–447.

Moss, D.W., Campbell, D.M., Anagnostou-Kakaras, E. and King, E.J. (1961a), *Pure Appl. Chem.*, **3**, 397–402.

Moss, D.W., Eaton, R.H., Smith, J.K. and Whitby, L.G. (1966), *Biochem. J.*, **98**, 32C–33C.

Moss, D.W., Eaton, R.H., Smith, J.K. and Whitby, L.G. (1967), *Biochem. J.*, **102**, 53–57.

Moss, D.W. and King, E.J. (1962), *Biochem. J.*, **84**, 192–195.

Moss, D.W., Shakespeare, M.J. and Thomas, D.M. (1972), *Clin. Chim. Acta*, **40**, 35–41.

Moss, D.W. and Walli, A.K. (1969), *Biochim. Biophys. Acta*, **191**, 476–477.

Moss, D.W. and Whitby, L.G. (1975), *Clin. Chim. Acta*, **61**, 63–71.

Muensch, H., Yoshida, A., Altland, K., Jensen, W. and Goedde, H.-W. (1978), *Amer. J. Hum. Genet.*, **30**, 302–307.

Nadal-Ginard, B. and Markert, C.L. (1975), In *Isozymes*, Vol. 2, (Ed Markert, C.L.), Academic Press, New York, pp. 45–67.

Nadler, H.L. and Egan, T.J. (1970), *New Engl. J. Med.*, **282**, 302–307.

Nagamine, M. (1972), *Clin. Chim. Acta*, **36**, 139–144.

Nagamine, M. and Ohkuma, S. (1975), *Clin. Chim. Acta*, **65**, 39–46.

Nakayama, T., Yoshida, M. and Kitamura, M. (1970), *Clin. Chim. Acta*, **30**, 546–548.

Nance, W.E., Claflin, A. and Smithies, O. (1963), *Science*, **142**, 1075–1077.

Nayudu, P.R.V. and Hercus, F.B. (1974), *Biochem. J.*, **141**, 93–101.

Nealon, D.A. and Henderson, A.R. (1975), *J. Clin. Pathol.*, **28**, 834–836.

Neilands, J.B. (1952), *Science*, **115**, 143–144.

Neumeier, D., Prellwitz, W., Würzburg, U., Brundobler, M., Olbermann, M., Just, H.-J., Knedel, M. and Lang, H. (1976), *Clin. Chim. Acta*, **73**, 445–451.

Nisselbaum, J.S. and Bodansky, O. (1959), *J. Biol. Chem.*, **234**, 3276–3280.

Norden, A.G.W., Tennant, L.L. and O'Brien, J.S. (1974), *J. Biol. Chem.*, **249**, 7969–7976.

Nørgaard-Pedersen, N. (1973), *Scand. J. Immunol.* **2**, Suppl. 1, 125–128.

Notstrand, B., Vaara, I. and Kannan, K.K. (1975), In *Isozymes*, Vol. 1, (Ed. Markert, C.L.), Academic Press, New York, pp. 575–599.

O'Brien, J.S., Okada, S., Fillerup, D.L., Veath, M.L., Adornato, B., Brenner, P.H. and Leroy, J.G. (1971), *Science*, **172**, 61–64.

Ogawa, M., Kosaki, G., Matsuura, K., Fujimoto, K.-I., Minamiura, N., Yamamoto, T. and Kikuchi, M. (1978), *Clin. Chim. Acta*, **87**, 17–21.

Ornstein, L. (1964), *Ann. NY Acad. Sci.*, **121**, 321–349.

Panveliwalla, D.K. and Moss, D.W. (1966), *Biochem. J.*, **99**, 501–506.

Pearce, J., Edwards, Y.H. and Harris, H. (1976), *Ann. Hum. Genet.*, **39**, 263–276.

Peacock, A.C., Reed, R.A. and Highsmith, E.M. (1963), *Clin. Chim. Acta*, **8**, 914–917.

Penhoet, E., Rajkumar, T. and Rutter, W.J. (1966), *Proc. Natl. Acad. Sci. USA*, **56**, 1275–1282.

Pentchev, P.G., Brady, R.O., Hibbert, S.R., Gal, A.E. and Shapiro, D. (1973), *J. Biol. Chem.*, **248**, 5256–5261.

Penton, E., Poenaru, L. and Dreyfus, J.C. (1975), *Biochim. Biophys. Acta*, **391**, 162–169.

Pesce, A., Fondy, T.P., Stolzenbach, A., Castillo, F. and Kaplan, N.O. (1967), *J. Biol. Chem.*, **242**, 2151–2167.

Peterson, J.S., Chern, C.J., Hawkins, R.N. and Black, J.A. (1974), *FEBS Letters*, **39**, 73–77.

PetitClerc, C. (1976), *Clin. Chem.*, **22**, 42–48.

Pfleiderer, G., Dikow, A.L. and Falkenberg, F. (1974), *Hoppe-Seyler's Z. Physiol. Chem.*, **355**, 223–228.

Pfleiderer, G. and Jeckel, D. (1957), *Biochem. Z.*, **329**, 370–380.

Pfleiderer, G. and Wachsmuth, E.D. (1961), *Biochem. Z.*, **334**, 185–198.

Philip, J. and Vesell, E.S. (1962), *Proc. Soc. Exp. Biol. Med.*, **110**, 582–585.

Phillips, J.P., Jones, H.M., Hitchcock, R., Adams, N. and Thompson, R.J. (1980), *Brit. Med. J.*, **281**, 777–779.

Phillips, N.C., Robinson, D. and Winchester, B. (1975), *Biochem. J.*, **151**, 469–475.

Plagemann, P.G.W., Gregory, K.F. and Wroblewski, F. (1960), *J. Biol. Chem.*, **235**, 2288–2293.

Plagemann, P.G.W., Gregory, K.F. and Wroblewski, F. (1961), *Biochem. Z.*, **334**, 37–48.

Plocke, D.J. and Vallee, B.L. (1962), *Biochemistry*, **1**, 1039–1043.

Plummer, D.T., Elliott, B.A., Cooke, K.B. and Wilkinson, J.H. (1963), *Biochem. J.*, **87**, 416–422.

Plummer, D.T. and Wilkinson, J.H. (1963), *Biochem. J.*, **87**, 423–429.

Podolsky, D.K. and Weiser, M.M. (1980), *J. Biol. Chem.*, **254**, 3983–3990.

Podolsky, D.K., Weiser, M.M., Isselbacher, K.J. and Cohen, A.M. (1978), *New. Engl. J. Med.*, **299**, 703–705.

Poenaru, L. and Dreyfus, J.C. (1973), *Clin. Chim. Acta*, **43**, 439–442.

Polin, S.G., Spellberg, M.A., Teitelman, L. and Okumura, M. (1962), *Gastroenterol.*, **42**, 431–438.

Purich, D.L., Fromm, H.J. and Rudolph, F.B. (1973), *Adv. Enzymol.*, **39**, 249–326.

Raftell, M. and Blomberg, F. (1974), *Europ. J. Biochem.*, **49**, 31–39.

Rao, P.S., Lukes, J.J., Ayres, S.M. and Mueller, H. (1975), *Clin. Chem.*, **21**, 1612–1618.

Reinitz, G.L. (1977), *Biochem. Genet.*, **15**, 445–454.

Revis, N.W., Thomson, R.Y. and Cameron, A.J.V. (1977), *Cardiovascular Res.*, **11**, 172–176.

Richterich, R., Schafroth, P. and Aebi, H. (1963), *Clin. Chim. Acta*, **8**, 178–192.

Richterich, R., Schafroth, P. and Franz, H.E. (1961), *Enzymol. Biol. Clin.*, **1**, 114–122.

Rider, C.C. and Taylor, C.B. (1974), *Biochim. Biophys. Acta*, **365**, 285–300.

Rider, C.C. and Taylor, C.B. (1975), *Biochim. Biophys. Acta*, **405**, 175–187.

Rider, C.C. and Taylor, C.B. (1976), *Biochim. Biophys. Acta*, **452**, 245–252.

Righetti, A.B.B. and Kaplan, M.M. (1974), *Proc. Soc. Exp. Biol. Med.*, **145**, 726–728.

Roberts, R., Sobel, B.E. and Parker, C.W. (1976), *Science*, **194**, 855–857.

Robinson, D. and Stirling, J.L. (1968), *Biochem. J.*, **107**, 321–327.

Robinson, D.B. and Glew, R.H. (1980), *Clin, Chem.*, **26**, 371–382.

Robinson, J.C. and Pierce, J.E. (1964), *Nature*, **204**, 472–473.

Robson, E.B. and Harris, H. (1966), *Ann. Hum. Genet.*, **30**, 219–232.

Romslo, I., Bjark, P. and Solberg, P.O. (1971), *Scand. J. Clin. Lab. Invest.*, **28**, 21–26.

Rosalki, S.B. (1965), *Nature*, **207**, 414.

Rosalki, S.B. (1968), In *Proceedings of the Fourth Symposium on Current Research in Muscular Dystrophy*, Pitman Medical, London, pp. 348–357.

Rosalki, S.B. and Wilkinson, J.H. (1960), *Nature*, **188**, 1110–1111.

Roy, A.V., Brower, M.E. and Hayden, J.E. (1971), *Clin. Chem.*, **17**, 1093–1102.

Rubin, C.S., Dancis, J., Yip, L.C., Nowinsky, R.C. and Balis, M.E. (1971), *Proc. Natl. Acad. Sci. USA*, **68**, 1461–1464.

Ryan, J.P., Appert, H.E., Carballo, J. and Davies, R.H. (1975), *Proc. Soc. Exp. Biol. Med.*, **148**, 194–197

Saks, V.A., Lipina, N.B., Sharov, V.G., Smirnov, V.N., Chazov, E. and Grosse, R. (1977), *Biochim. Biophys. Acta*, **465**, 550–558.

Salthe, S.N., Chilson, O.P. and Kaplan, N.O. (1965), *Nature*, **207**, 723–726.

Samuelson, R.C. and Moss, D.W. (1978), *Clin. Chim. Acta*, **83**, 167–170.

Scandalios, J.G. (1974), *Annu. Rev. Plant Physiol.*, **25**, 225–258.

Scandalios, J.G. (1975), In *Isozymes*, Vol. 3, (Ed. Markert, C.L.), Academic Press, New York, pp. 213–238.

Schapira, F., Dreyfus, J.-C., Allard, D. and Gregon-Lauer, C. (1968), *Clin. Chim. Acta*, **20**, 439–447.

Schapira, F., Hatzfeld, A. and Weber, A. (1975), In *Isozymes*, Vol. 3, (Ed Markert, C.L.), Academic Press, New York, pp. 987–1003.

Schechter, A.J. and Epstein, C.J. (1968), *Science*, **159**, 997–999.

Schlaeger, R. (1975), *Z. Klin. Chem. Klin. Biochem.*, **13**, 277–281.

Schlamowitz, M., and Bodansky, O. (1959), *J. Biol. Chem.*, **234**, 1433–1437.

Schlesinger, M.J., Bloch, W. and Kelley, P.M. (1975), In *Isozymes*, Vol. 1, (Ed. Markert, C.L.), Academic Press, New York, pp. 333–342.

Schmechel, D., Marangos, P.J., Zis, A.P., Brightman, M. and Goodwin, F.K. (1978), *Science*, **199**, 313–315.

Schmechel, D., Marangos, P.J. and Brightman, M. (1978a), *Nature*, **276**, 834–836.

Schmidt, E., Schmidt, F.W. and Otto P. (1967), *Clin. Chim. Acta*, **15**, 283–289.

Schoenenberger, G.A., Buser, S., Cueni, L., Döbeli, H., Gillessen, D., Lergier, W., Schöttli, G., Tobler, H.J. and Wilson, K. (1980), *Regulatory Peptides*, **1**, 223–244.

Scholl, A. and Eppenberger, H.M. (1969), *Experientia*, **25**, 794–796.

Schwartz, J.H., Crestfield, A.M. and Lipmann, F. (1963), *Proc. Natl. Acad. Sci. USA*, **49**, 722–728.

Scott, E.M. and Powers, R.F. (1972), *Nature New Biol.*, **236**, 83–84.

Scutt, P.B. and Moss, D.W. (1968), *Enzymologia*, **35**, 157–167.

Sebesta, D.G., Bradshaw, F.J. and Prockop, D.J. (1964), *Gastroenterol.*, **47**, 166–170.

Shaklee, J.B., Kepes, K.L. and Whitt, F.S. (1973), *J. Exp. Zool.*, **185**, 217–240.

Shapiro, A.L., Vinuela, E. and Maizel, J.V., Jr. (1967), *Biochem. Biophys. Res. Commun.*, **28**, 815–820.

Shaw, C.R. and Prasad, R. (1970), *Biochem. Genet.*, **4**, 297–320.

Shibuta, Y., Higashi, T., Hirai, H. and Hamilton, H.B. (1967), *Arch. Biochem. Biophys.*, **118**, 200–209.

Shinkai, K. and Akedo, H. (1972), *Cancer Res.*, **32**, 2307–2313.

Shows, T.B. (1977), *Isozymes: Current Topics in Biological and Medical Research*, **2**, 107–158.

Shows, T.B., Chapman, V.M. and Ruddle, F.H. (1970), *Biochem. Genet.*, **4**, 707–718.

Skude, G. and Kollberg, H. (1976), *Acta Paediat. Scand.*, **65**, 145–149.

Singer, R.M. and Fishman, W.H. (1975), In *Isozymes*, (Ed. Markert, C.L.), Academic Press, New York, pp. 753–774.

Slaughter, C.A., Hopkinson, D.A. and Harris, H. (1977), *Ann. Hum. Genet.*, **40**, 385–401.

Smith, A.F., Radford, D., Wong, C.P. and Oliver, M.F. (1976), *Brit. Heart. J.*, **38**, 225–232.

Smith, I., Lightstone, P.J. and Perry, J.D. (1971), *Clin. Chim. Acta*, **35**, 59–66.

Smith, J.K., Eaton, R.H., Whitby, L.G. and Moss, D.W. (1968), *Analyt. Biochem.*, **23**, 84–96.

Smith, J.K. and Moss, D.W. (1968), *Analyt. Biochem.*, **25**, 500–509.

Smith, J.K. and Whitby, L.G. (1968), *Biochim. Biophys. Acta*, **151**, 607–618.

Smith, M., Hopkinson, D.A. and Harris, H. (1971a), *Ann. Hum. Genet.*, **34**, 251–271.

Smithies, O. (1955), *Biochem. J.*, **61**, 629–641.

Snaith, S.M. (1977), *Biochem. J.*, **163**, 557–564.

Somer, H., Donner, M., Murros, J and Konttinen, A. (1973), *Arch. Neurol.*, **29**, 343–345.

Somer, H. and Konttinen, A. (1972), *Clin. Chim. Acta*, **40**, 133–138.

Spencer, C.J. and Gelehrter, T.D. (1974), *J. Biol. Chem.*, **249**, 577–583.

Srivastava, S.K., Ansari, N.H., Hawkins, L.A. and Wiktorowicz, J.E. (1979), *Biochem. J.*, **179**, 657–664.

Srivastava, S.K., Yoshida, A., Awasthi, Y.C. and Beutler, E. (1974), *J. Biol. Chem.*, **249**, 2049–2053.

Stadtman, E.R. (1968), *Ann. NY Acad. Sci.*, **151**, 516–530.

Stagg, B.H. and Whyley, G.A. (1968), *Clin. Chim. Acta*, **19**, 139–145.

Stanbury, J.B., Wyngaarden, J.B. and Fredrickson, D.S. (1978), *The Metabolic Basis of Inherited Disease*, 4th Edn, McGraw-Hill, New York.

Sugiura, M. and Hirano, K. (1977), Biochem. Med., **17**, 222–228.

Sur, B.K., Moss, D.W. and King, E.J. (1962), *Proc. Assoc. Clin. Biochemists*, **2**, 11–13.

Sutton, H.E. and Wagner, R.P. (1975), *Annu. Rev. Hum. Genet.*, **9**, 187–212.

Swallow, D. and Harris, H. (1972), *Ann. Hum. Genet.*, **36**, 141–152.

Takahashi, K., Shutta, K., Matsuo, B., Takai, T., Takao, H. and Imura, H. (1977), *Clin. Chim. Acta*, **75**, 435–442.

Takeuchi, T., Matsushima, T., Sugimura, T., Kozu, T., Takeuchi, T. and Takemoto, T. (1974), *Clin. Chim. Acta*, **54**, 137–144.

Tapia, F.J., Barbosa, A.J.A., Marangos, P.J., Polak, J.M., Bloom, S.R., Dermosly, C. and Pearse, A.G.E. (1981), *Lancet,* i, 808–811.

Tashian, R.E., Riggs, S.K. and Ya-Shiou, L.Yu. (1966), *Arch. Biochem., Biophys.*, **117**, 320–327.

Tashian, R.E. (1977), *Isozymes: Current Topics in Biological and Medical Research*, **2**, 21–62.

Tedesco, T.A. (1972), *J. Biol. Chem.*, **247**, 6631–6636.

Teller, D.C. (1976), *Nature*, **260**, 729–731.

Thomas, D.M. and Moss, D.W. (1972), *Enzymologia*, **42**, 65–77.

Thomson, R.J., Rubery, E.D. and Jones, H.M. (1980), *Lancet*, ii, 673–675.

Thompson, S.T. and Stellwagen, E. (1976), *Proc. Natl. Acad. Sci. USA*, **73**, 361–365.

Timperley, W.R. (1968), *Lancet*, ii, 356.

Timperley, W.R., Turner, P and Davies, S, (1971), *J. Pathol.*, **103**, 257–262.

Tolley, E. and Craig, I. (1975), *Biochem. Genet.*, **13**, 867–883.

Tsai, M.Y., Gonzalez, F. and Kemp, R.G. (1975), In *Isozymes*, Vol. 2, (Ed. Markert, C.L.), Academic Press, New York, pp. 819–835.

Tsai, M.Y. and Kemp, R.G. (1972), *Arch. Biochem., Biophys.*, **150**, 407–411.

Turner, B.M., Fisher, R.A. and Harris, H. (1974), *Ann. Hum. Genet.*, **37**, 455–467.

Tzvetanova, E. (1971), *Enzyme*, **12**, 279–288.

Uriel, J. (1963), *Ann. NY Acad. Sci.*, **103**, 956–979.

Usategui-Gomez, M., Wicks, R.W., Farrenkopf, B., Hager, H. and Warshaw, M. (1981), *Clin. Chem.*, **27**, 823–827.

Van Belle, H. (1976), *Clin. Chem.*, **22**, 972–976.

Van Berkel, J.V., Kosler, J.F. and Hulsmann, W.C. (1972), *Biochim. Biophys. Acta*, **276**, 425–429.

Van der Helm, H.J. (1962), *Clin. Chim. Acta*, **7**, 124–128.

Van Etten, R.L. and Saini, M.S. (1978), *Clin. Chem.*, **24**, 1525–1530.

Van Lente, F. and Galen, R.S. (1978), *Clin. Chim. Acta*, **87**, 211–217.

Van Noorden, S. and Polak, J.M. (1981). In *Gut Hormones*, (Ed. Bloom, S.R. and Polak, J.M.), Churchill Livingstone, Edinburgh, pp. 80–89.

Van Wijhe, M., Blanchaer, M.C. and St. George-Stubbs, S. (1964), *J. Histochem. Cytochem.*, **12**, 608–614.

Vesell, E.S. and Bearn, A.G. (1961), *J. Clin. Invest.*, **60**, 586–591.

Vessell, E.S. and Pool, P.E. (1966), *Proc. Natl. Acad. Sci. USA*, **55**, 756–762,

Vesell, E.S. and Yielding, K.L. (1968), *Ann. NY Acad. Sci.*, **151**, 678–689.

Vihko, P. (1978), *Clin. Chem.*, **24**, 1783–1787.

Vihko, P., Kontturi, M. and Kovhonen, L.K. (1978), *Clin. Chem.*, **24**, 466–470.

Vihko, P., Sajanti, E., Janne, O., Peltonen, L. and Vihko, R. (1978a), *Clin. Chem.*, **24**, 1915–1919.

Vladutiu, A.O., Schachner, A., Schaeffer, P.A., Schimert, G., Lajos, T.Z., Lee, A.B. and Siegel, J.H. (1977), *Clin. Chim. Acta*, **75**, 467–473.

Vora, S., Seaman, C., Durham, C. and Piomelli, S. (1980), *Proc. Natl. Acad. Sci. USA*, **77**, 62–66.

Wada, H. and Morino, Y. (1964), *Vitam. Horm.*, **22**, 411–444.

Wagner, G.S., Roe, C.R., Limbird, L.E., Rosati, R.A. and Wallace, A.G. (1973), *Circulation*, **47**, 263–269.

Walker, P.R. (1974), *Life Sciences*, **24**, 89–96.

Walker, A.W. (1974a), *Clin. Chim. Acta.*, **55**, 399–405.

Walter, H., Selby, F.W. and Francisco, J.R. (1965), *Nature*, **208**, 76–77.

Walton, R.J., Preston, C.J., Russell, R.G.G. and Kanis, J.A. (1975), *Clin. Chim. Acta*, **63**, 227–229.

Warnock, M.L. and Reisman, R. (1969), *Clin. Chim. Acta*, **24**, 5–11.

Warshaw, A.L. (1977), *J. Lab. Clin. Med.*, **90**, 1–3.

Watson, R.A. and Tang, D.B. (1980), *New Engl. J. Med.*, **303**, 497–499.

Watts, D.C. (1968), *Adv. Comp. Biochem. Physiol.*, **3**, 1–114.

Watts, D.C., Foçant, B., Moreland, B.M. and Watts, R.L. (1972), *Nature New Biol.*, **237**, 51–53.

Werthamer, S., Frieberg, A. and Amaral, L. (1973), *Clin. Chim. Acta*, **45**, 5–8.

Wevers, R.A., Olthuis, H.P., Van Niel, J.C.C., Van Wilgenburg, M.G.M. and Soons, J.B.J. (1977), *Clin. Chim. Acta*, **75**, 377–385.

Wevers, R.A., Wolters, R.J. and Soons, J.B.J. (1977a), *Clin. Chim. Acta*, **78**, 271–276.

Whitaker, K.B. and Moss D.W. (1979), *Biochem. J.*, **183**, 189–192.

Whitaker, K.B., Eckland, D., Hodgson, H.J.F., Saverymuttu, S., Williams, G. and Moss, D.W. (1982), *Clin Chem.*, in press.

Whitaker, K.B., Whitby, L.G. and Moss, D.W. (1977), *Clin. Chim. Acta*, **80**, 209–220.

Whitaker, K.B., Whitby, L.G. and Moss, D.W. (1978), In *Enzymes in Health and Disease*, Karger, Basel, pp. 127–130.

Whitby, L.G. and Moss, D.W. (1975), *Clin. Chim. Acta*, **59**, 361–367.

Whitelaw, A.G.L., Rogers, P.A., Hopkinson, D.A., Gordon, H., Emerson, P.A., Darley, J.H., Reid, C. and Crawfurd, M.d'A. (1979), *J. Med. Genet.*, **16**, 189–196.

Whitt, G.S. (1969), *Science*, **166**, 1156–1158.

Wieland, T., Georgopoulos, D., Kampe, H. and Wachsmuth, E.D. (1964), *Biochem. Z.*, **340**, 483–486.

Wieland, T. and Pfleiderer, G. (1957), *Biochem. Z.*, **329**, 112–116.

Wieland, T. and Pfleiderer, G. (1962), *Angew. Chem. Internatl.* Edn., **1**, 169–224.

Wieland, T., Pfleiderer, G. and Ortanderl, F. (1959), *Biochem. Z.*, **331**, 103–109.

Wieme, R.J. (1959), *Clin. Chim. Acta*, **4**, 46–50.

Wieme, R.J. and Herpol, J.W. (1962), *Nature*, **194**, 287–288.

Wilhelm, A. (1979), In *Clincial Enzymology Symposia*, Vol. 2, (Ed. Burlina, A. and Galzinga, L.), Piccin Medical, Padua, pp. 467–473.

Wilkinson, J.H. and Qureshi, A.R. (1976), *Clin. Chem.*, **22**, 1269–1276.

Wilson, A.C., Cahn, R.D. and Kaplan, N.O. (1963), *Nature*, **197**, 331–334.

Willson, V.J.C., Jones, H.M. and Thompson, R.J. (1981), *Clin. Chim. Acta*, **113**, 153–163.

Witteveen, S.A.J.G., Sobel, B.E. and DeLuca, M. (1974), *Proc. Natl. Acad. Sci. USA*, **71**, 1384–1387.

Wolf, R.O., Taussig, L.M., Ross, M.E. and Wood, R.E. (1976), *J. Lab. Clin. Med.*, **87**, 164–168.

Wootton, A.M., Neale, G. and Moss, D.W. (1977), *Clin. Sci. Mol. Med.*, **52**, 585–590.

Wroblewski, F. and Gregory, K.F. (1961), *Ann. NY Acad. Sci.*, **94**, 912–932.

Yam, L.T. (1974), *Amer. J. Med.*, **56**, 604–616.

Yang, N.-S. (1975), In *Isozymes*, Vol. 3, (Ed. Markert, C.L.), Academic Press, New York, pp. 191–212.

Yoshida, A. (1967), *Proc. Natl. Acad. Sci. USA*, **57**, 835–840.

Yoshida, A., Steinmann, L. and Harbart, P. (1967), *Nature*, **216**, 275–276.

Zinkham, W.H., Blanco, A. and Kupchyk, L. (1963), *Science*, **142**, 1303–1304.

Zinkham, W.H., Inensee, H. and Renwick, J.H. (1969), *Science*, **164**, 185–187.

Zondag, H.A. and Klein, F. (1968), *Ann. NY Acad. Sci.*, **151**, 578–586.

Zweig, M.H., Van Steirteghem, A.C. and Schechter, A.N. (1978), *Clin. Chem.*, **24**, 422–428.

Index